Biochemistry: A Very Short Introduction

VERY SHORT INTRODUCTIONS are for anyone wanting a stimulating and accessible way into a new subject. They are written by experts, and have been translated into more than 45 different languages.

The series began in 1995, and now covers a wide variety of topics in every discipline. The VSI library currently contains over 650 volumes—a Very Short Introduction to everything from Psychology and Philosophy of Science to American History and Relativity—and continues to grow in every subject area.

Very Short Introductions available now:

WORLD MUSIC Philip Bohlman
THE WORLD TRADE
   ORGANIZATION
   Amrita Narlikar

WORLD WAR II Gerhard L. Weinberg
WRITING AND SCRIPT
   Andrew Robinson
ZIONISM Michael Stanislawski

Available soon:

HUMAN PHYSIOLOGY
   Jamie A. Davies
PAKISTAN Pippa Virdee
HORROR Darryl Jones

DIPLOMATIC HISTORY
   Joseph M. Siracusa
BLASPHEMY
   Yvonne Sherwood

For more information visit our website

www.oup.com/vsi/

Mark Lorch

BIOCHEMISTRY

A Very Short Introduction

OXFORD
UNIVERSITY PRESS

OXFORD
UNIVERSITY PRESS

Great Clarendon Street, Oxford, OX2 6DP,
United Kingdom

Oxford University Press is a department of the University of Oxford.
It furthers the University's objective of excellence in research, scholarship,
and education by publishing worldwide. Oxford is a registered trade mark of
Oxford University Press in the UK and in certain other countries

Published in the United States of America by Oxford University Press
198 Madison Avenue, New York, NY 10016, United States of America

British Library Cataloguing in Publication Data
Data available

Library of Congress Control Number: 2020949465

ISBN 978–0–19–883387–1

Printed and bound by
CPI Group (UK) Ltd, Croydon, CR0 4YY

Links to third party websites are provided by Oxford in good faith and
for information only. Oxford disclaims any responsibility for the materials
contained in any third party website referenced in this work.

Contents

Preface

From the simplest bacteria to humans, all living things are composed of cells of one type or another. Amazingly, no matter where on the evolutionary tree they perch, those organisms all have fundamentally the same chemistry. This chemistry must provide mechanisms that allow cells to interact with the external world, a means to power the cell, machinery to carry out all the varied processes, a structure within which everything runs, and of course some sort of governance. Cells, in many ways, are like communities, but controlled and governed through a web of interlocking chemical reactions. Biochemistry is the study of those reactions, the molecules that are created, manipulated, and destroyed as a result of them, and the massive macromolecules (such as DNA, cytoskeletons, proteins, and carbohydrates) that form the chemical machinery and structures on which these biochemical reactions take place.

Or, put more succinctly by the great physicist Erwin Schrödinger, 'In biology...a single group of atoms...produces orderly events marvellously tuned in with each other and the environment according to the most subtle laws.'

Biochemistry is then the endeavour to understand those subtle laws governing those finely tuned orderly events; it is the study of biological molecules and their interactions, and so aims to reveal the molecular basis of life.

Of course, life in all its glory is made up of so much more than just single cells. Cells come together to form multi-cellular organisms which then require a means for individual cells to communicate and 'trade' with one another. The organisms, in turn, interact to form the complex webs that are our eco-systems. And all of those interactions are modulated and facilitated through biochemical means. For example, consider the rhodopsin molecules that respond to photons of light, and so act as the first stage of a predator spotting its next meal. Or the olfactory proteins that bind a few minuscule molecules, which trigger a cascade of biochemical reactions that result in prey being alerted to the predator's presence. Or the antibodies that act as the first guards, recognizing the foreign molecules of an invading parasite and triggering the army that is the immune response. All of these processes fall within the realm of biochemistry.

It didn't take long for an understanding of the chemistry of life to turn into a desire to manipulate it. Drugs and therapies are all aimed at modifying biochemical processes for good or ill, so that penicillin, for example, derived from a particular mould, stops bacteria making their cell walls; aspirin, with its origins in willow bark, inhibits the enzymes involved in inflammatory responses; a few nanograms of botulinum toxin (botox) can kill by preventing the release of neurotransmitters from the ends of nerves, leading to paralysis and death—or, alternatively, the same botulinum toxin administered in tiny quantities results in a wrinkle-free forehead. This is all biochemistry.

Detailed description of these topics could easily have made it into this book, and some readers may feel I was remiss in neglecting them along with other topics as fundamental as vitamins, hormones, chromosomes, and numerous biochemical techniques. But this is after all a very short introduction, and so I had to draw the line somewhere. As a result, for much of the book I've focused on some of the chemistry that occurs within cells. For therein lie the fundamental chemical processes that all life shares.

Finally, the boundaries of biochemistry are ill defined, overlapping with genetics, molecular biology, cell biology, biophysics, and biotechnology. This being so, I round off the volume with two chapters exploring some of the fundamental discoveries in biochemistry that are influencing these other fields and society generally.

List of illustrations

Chapter 1
The roots of biochemistry

In many ways the history of biochemistry is linked to the understanding of arguably the oldest uses of biotechnology—fermentation and the production of alcoholic beverages and cheese. Cheese manufacturing probably coincides with the development of agriculture and domestication of goats and sheep some 10,000 years ago. The fresh milk, left in warm conditions, would have been a perfect breeding ground for microbes such as *Lactococcus lactis*. These bacteria convert the lactose in the milk to lactic acid, which in turn cause proteins in the milk to coagulate, forming curds. The resulting reduction in the lactose content of the curds (and left-over whey) would have been a great boon to adult Neolithic humans as they undoubtedly had trouble digesting the lactose (not yet having evolved a means to break down this sugar). And of course, the curds can be pressed and dried to form readily storable cheese. Around the same time other cultures made use of yeast's ability to ferment glucose into ethanol. In the process, the yeast outcompeted harmful bacteria, so rendering the resulting beverage safer to drink than untreated water. And so, beer was born.

Fermentation and enzymes

Fast forward to the 19th century and an argument was brewing between two titans of the scientific community over the chemical processes underpinning fermentation. Louis Pasteur, the microbiologist and chemist, more famous for his work on vaccination and pasteurization, clashed with Justus Freiherr von Liebig (widely known for the condenser named after him, but more importantly as the founder of the fertilization industry and father of organic chemistry).

Liebig supported the chemical theory of fermentation. He built on ideas proposed by Antoine Lavoisier that yeast accompanied and possibly triggered the onset of fermentation, but once the biochemical ball was rolling, living organisms took no further part in the process. Liebig explained this by theorizing that fermentation was a result of transfer of molecular instabilities between the components of the ferment, a process that he thought was comparable to putrefaction of animal and vegetable matter. Hence, from his viewpoint, fermentation was a purely chemical process that did not require any intervention from living organisms.

In contrast, Pasteur took an experimental approach. By showing that fermentation did not take place in a broth where micro-organisms had been killed by boiling, Pasteur concluded that fermentation was due to the action of viable micro-organisms. In his words, 'alcoholic fermentation is an act correlated with the life and organization of the yeast cells, not with the death or putrefaction of the cells'.

This statement was clearly a dig at Liebig's theories, and helped to fuel the argument between the two scientists, which took place via their papers and letters throughout the late 1850s and into the 1860s. At times their correspondence was quite bitter, with Liebig suggesting that Pasteur's experiments could not be replicated. In truth neither's theories on fermentation were quite correct. Ultimately, the matter was settled some thirty years later by the German brothers, Hans Büchner (a bacteriologist) and Eduard Büchner (a chemist).

Whilst the fermentation debate was raging, the field of enzymology began to blossom. It had been known for some time that living organisms excreted molecules that allowed chemical reactions to proceed much faster and under milder conditions than would otherwise be expected. This had been shown first by Anselme Payen (in 1833), who isolated a chemical from malt extract that converted starch to sugar (he called it diastase; we now refer to it as amylase). Shortly afterwards, in 1834, Theodor Schwann extracted pepsin (from the Greek πέψις, or *pepsis*, meaning 'digestion') from gastric juices. Pepsin broke down egg proteins much faster than stomach acid alone. Others followed and soon found more examples of extracts, all of which degraded various biologically sourced chemicals. These materials clearly needed a name, and eventually Wilhelm Kühne, in 1876, provided one by coining the term enzyme (from the Greek ἔνζυμον, or *enzumous*, meaning 'in yeast/leaven'). It is important to note that at the time Kühne was quite clear that his definition of enzymes only covered materials found outside of cells.

In 1893, Hans Büchner was diligently working on a method to extract materials from bacteria for the purposes of immunization. Bacteria were difficult to grow in any quantity, so instead he tested his techniques on yeast (living in Munich with breweries aplenty, there was a very ready supply of yeast). Hans managed to break open cells and release their fluid interior—the protoplasm—by pulverizing yeast in a hydraulic press. After the process, the yeast was quite dead and so, according to Pasteur's theory, the extract should not support fermentation. Therefore the Büchners were surprised when, within a mixture of yeast protoplasm and glucose, carbon dioxide (CO_2) was spotted steadily bubbling to the surface of the reaction vessel. This tell-tale sign unequivocally demonstrated that fermentation could take place without a viable living agent.

The key to the Büchners' serendipitous discovery was the way Hans extracted the contents of the cells. Many others had attempted to extract active protoplasm, but they had all used heat or solvents, both of which kill the cells, but also destroy the delicate proteins within them. And it was these proteins, and specifically the enzymes *inside* the cells, that were responsible for fermentation. For the first time, the Büchners had shown that chemicals extracted from biological cells carried out the same actions both inside and outside the cell. By extracting the zymase (as Eduard Büchner called it), they demonstrated that fermentation could occur without the presence of a living organism, but did require material derived from cells. In essence, they had settled on a middle ground between Liebig and Pasteur's positions. Moreover, they had widened Wilhelm Kühne's definition of an enzyme to include material within protoplasm. This really was a revolutionary finding, and the nail in the coffin for vitalism, the widely held belief at the time that living things are fundamentally different (due to some non-physical agent) from non-living matter.

Proteins

By unshackling biochemistry (first used in the sense we use it today by Felix Hoppe-Seyler in 1877) from vitalism, the Büchners demonstrated that the inner workings of cells could be studied using the same techniques applied to the rest of chemistry. This paved the way for modern biochemistry and unlocked the path to understanding the nature of enzymes, for it was still unclear that they were in fact proteins.

Indeed, at the same time as debates on fermentation and enzymology flourished, the nature of proteins was also under scrutiny. By the 1830s chemists were busy applying one of the few analytical techniques available to them—elemental analysis—to biological materials. Elemental analysis told the industrious researchers how much carbon, nitrogen, oxygen, hydrogen, phosphorus, sulfur, and so on, there was in the material they chose to investigate. Since there was not much else to do with this technique, scientists used it to analyse just about every sort of naturally occurring material they could lay their hands on. They produced huge tables of data detailing the elemental composition of plant materials including green tea, coffee, black tea, sugar (they seemed quite keen on analysing beverages), bark, and oil, to name a few examples. Their results gave them the percentage of all the elements in a sample. For example, common sugar (sucrose) worked out to be 27 per cent carbon, 49 per cent hydrogen, and 24 per cent oxygen. Today we write this as an empirical formula of $C_{12}H_{22}O_{11}$, meaning carbon, hydrogen, and oxygen are present at a ratio of 12:22:11. Pretty much everything that these early biochemists tested gave ratios of elements with numbers in the units, tens, or twenties.

Then in 1838 a young physician turned university lecturer, named Gerrit Mulder, switched his attention from plant to animal samples. In his Rotterdam lab he set his band of students the task of analysing egg white and muscle tissue. His findings were totally unexpected: the ratios of carbon, hydrogen, nitrogen, and oxygen were precisely the same for the two materials. Moreover, both contained tiny amounts of phosphorus and sulfur, again in the same ratios. This was astonishing for it suggested that extracts from entirely different sources, muscle and egg white, appeared to be identical in composition. Even more remarkably, the empirical formula was found to be $C_{400}H_{620}N_{100}P_1S_1$. Up to this point all the formulae derived from other substances tested looked more or less like that for sugar ($C_{12}H_{22}O_{11}$) or alcohol ($C_2H_7O_1$), that is, relatively small numbers for each of the elements. This new formula meant that the molecules they were testing in muscle fibres and egg white were much bigger than anything that had been encountered before. Furthermore, when Mulder went on to test other plant and animal materials, like blood serum (blood with all the cells and platelets removed) or wheat albumin (the water-soluble parts of wheat grains), they too gave almost the same formula except with slightly different amounts of sulfur and phosphorus.

A very excited Mulder wrote to his mentor, Jöns Jacob Berzelius (a Swedish elder statesman of chemistry, who developed the chemical formula notation used today), to explain his results. Together they came to the conclusion that most animal matter was derived from plants. Animals eat the plants and then modify this 'Grundstoff', which roughly translates as a basic building block or chemical precursor. Berzelius thought they needed a better name and, in 1838, came up with 'proteios' (derived from the Greek for primary), which eventually became 'protein'.

Mulder's results caused a considerable stir in the growing field of biochemistry. Some big names jumped on the bandwagon, most notably Liebig, who set about writing a book on the subject. Meanwhile, others sought to replicate Mulder's experiments. It soon became apparent that Mulder's analysis of the newly named proteins was not quite right and, whilst they were undoubtedly very big molecules, there was a lot more variation in their formulae than he had claimed. By this time Liebig's book based on Mulder's original results had been published, and he was furious to find that the central tenet of the book, that proteins are animals' 'primary' nutritional agent, had been shown to be false. Nevertheless, Liebig's involvement, plus the ensuing spat with Mulder, enlivened the whole field and drew in a whole new group of scientists keen to determine what these newly discovered, enormous protein molecules were made of.

So, by the 19th century, biochemists (or physiological chemists as they were generally known) were aware of proteins and enzymes but whether enzymes were proteins was still far from clear. Two camps emerged. One argued that enzyme activity invariably coincided with the presence of proteins, so enzymes must be proteins. The other argued that proteins merely carried enzymes, and pointed to the fact that other non-proteinaceous materials could catalyse reactions in much the same way as enzymes. The crux of the 'carrier' argument came down to the simple observation that enzyme activity could be measured in the apparent absence of protein. In truth, this was easily explained by the fact that analytical techniques of the time were not sensitive enough to detect small amounts of protein, but because enzymes are such efficient catalysts their activity could be detected even when minuscule amounts were present. In the end, the matter was settled by what amounted to the reverse of the 'carrier' camp's argument. In the 1920s, James Sumner set about attempting to isolate an enzyme in its pure form. No one had yet achieved this feat, and indeed many biochemists of the time thought it was a ridiculous idea. Nevertheless, Sumner persevered, progressively isolating and purifying urease from jack beans. Then in 1926 his effort bore fruit. He final managed to coax the enzyme to pass the ultimate test of protein purity and coalesce into a crystal. Sumner's work conclusively demonstrated that the only possible source of the enzyme activity was protein, and so the debate about the nature of enzymes should really have been laid to rest. But people are rarely willing to let go of their theories so easily. And it took another highly pure crystallized enzyme, in this case pepsin, produced by John Northrop in 1929, to finally settle the argument.

We left the tale of chemical composition of proteins with the realization that they are huge molecules—macromolecules—largely composed of carbon, nitrogen, hydrogen, and oxygen plus a smattering of sulfur and phosphorus. Following Liebig's and Mulder's disagreement, scientists were keen to figure out, in more detail, what these newly named, enormous protein molecules were made of.

The next part of the story turned out to be quite straightforward. Proteins are easily degraded by acid. So, a whole host of researchers began to investigate what was left when various proteins reacted with hydrochloric acid. They soon noticed that there were several different molecules in the residues of these reactions, but that they all possessed both an acidic (-COOH) and an amino ($-NH_2$) group and so they become known as amino acids (Figure 1). (Incidentally, the term 'residue' stuck and to this day

1. Structure of the twenty most commonly occurring amino acids.

8

biochemists can be heard referring to residues when they mean amino acids.)

Following all the analyses, there proved to be twenty common, naturally occurring, proteogenic amino acids (there are in fact over 500 found in nature but only twenty are encoded in the genetic code, and generally found in proteins) and the difference between them is dependent on the form of the side chains. These vary in size, shape, and charge, from glycine, with a single hydrogen atom for a side chain, to the biggest amino acid, tryptophan, which has a double carbon ring. Some have acidic side chains (aspartate and glutamate), while others are basic (arginine and lysine); many are oily and hydrophobic (like leucine and valine). And there's an oddity: proline, with its side chain that curls back and forms a ring with the main chain alpha-carbon. (Strictly speaking, this makes proline an imino acid, but biochemists generally overlook this bit of chemical pedantry.) Only two of the twenty contain sulfur (hence the element's scarcity in the early analysis of proteins) and none contain phosphorus (it turns out that this is frequently added much later on in the protein manufacturing process).

The next problem to face biochemists was determining how the amino acids are linked together to form proteins. This conundrum was solved simultaneously, but with two very different approaches, by Emil Fischer and Franz Hofmeister. Coincidentally they presented their results at the same conference in Karlsbad, in the present-day Czech Republic, in 1902.

Hofmeister sorted out the problem with some excellent feats of deduction. From a chemist's perspective, there are numerous ways in which amino acids could plausibly be linked together to form large molecules, such as via carbon–carbon bonds, ether links, or nitrogen–carbon bonds. Hofmeister surveyed these possibilities and then worked out which linkages would result in the observed characteristics of proteins. For example, he reasoned that the acid groups in amino acids must be bound to something (and thus their acidic effects nullified), or solutions of proteins would be very acidic (which they are not). From these sorts of observations, Hofmeister figured out that the amino acids in proteins must be linked via the acid group bonding to an amino group. This method of linking one amino acid to another became known as a 'peptide bond' (Figure 2).

Meanwhile, Emil Fischer took a bottom up, experimental approach. He started with a variety of amino acids and then tried to link them together in such a way to produce something that would behave like a protein. Thus, via a totally different route, he came to exactly the same conclusion as Hofmeister. This result so fired Fischer's ambition that, in 1905, he was heard to say, 'My entire yearning is directed toward the first synthetic enzyme.' Quite an ambition, especially since this goal wasn't actually achieved until the 1990s.

The consequence of Hofmeister's and Fischer's discovery of the peptide bond is that proteins were understood to be long, linear chains (polymers) formed from amino acids, with one following the other. Short sequences of amino acids linked in this way became known simply as peptides, whilst longer chains were referred to as polypeptides. At this point it is worth noting that there is another way by which one of the amino acids frequently

Peptide bond

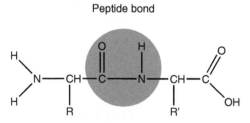

2. **The peptide bond, elucidated by Hofmeister and Fischer. The 'R' represents the side chains of amino acids.**

forms bonds within proteins. The sulfur at the end of cysteine side chains is capable of linking with one other to form disulfide bonds. This cross-linking adds some robustness to the protein structures, and so it is often found in structural proteins such as keratin, which makes up most of our hair.

Knowing that proteins are constructed of twenty different amino acids linked together in chains is important, but it doesn't actually tell you much about the protein itself. It's rather like being aware that sentences are made from twenty-six letters that appear one after another on a line, but that is of no help without another critical piece of information—the order of the letters. So it was with proteins. The next great challenge was to work out the amino acid sequence in a protein. The answer to this problem had to wait forty years and for a young PhD student, Frederick Sanger.

Sanger was, initially, just trying to develop a way to determine the first amino acid in a protein. He came up with a simple method that involved reacting a chemical called dinitrofluorobenzene (DNP) with the end of the protein molecule. When the DNP attached to the protein it turned yellow, and then Sanger put the composite in acid, which broke the protein up into its amino acids. Throughout this process the DNP usually stayed bound to the first amino acid in the chain. Next, he separated all the amino acids from one another using chromatography, then simply cut the yellow spot out of his chromatogram and analysed its contents.

Sometimes the product of the protein/DNP reaction was not particularly stable so when Sanger put the mix into acid the DNP became detached. To overcome this, he tried shorter incubations of protein in the acid. This seemed to do the job: the DNP stayed attached to the protein. But now the reaction was not long enough for the protein to be completely degraded into its constituent amino acids. This is when Sanger realized he had got lucky, because he noticed that now there were multiple yellow spots on his chromatography paper. He quickly concluded that the extra spots were due to bonds staying intact between the first and second amino acid, which meant he had the opportunity to figure out not only what the first amino acid was, but also the second. And by extending the method further, he could work out the whole protein sequence. However, while he had indeed found a way to sequence a protein, the actual work was still to be quite arduous. It took him another ten years to determine the amino acid sequence of one protein—insulin. Nevertheless, the work was seminal, for it demonstrated that the amino acids in any given protein were ordered in a very specific way. This order became known as the primary structure of proteins.

Then followed the question of how these strings of amino acids fold up into particular shapes. Answering this requires two more pieces of information: the shape of the amino acids plus knowledge of the attractive and repulsive forces between them. The latter came from the great chemist Linus Pauling. In his 1939 book *The Nature of the Chemical Bond*, he described how hydrogen bonds play an important part in determining the configuration of proteins. These weak electrostatic attractions occur when hydrogen atoms covalently bond to electronegative atoms (such as nitrogen and oxygen); this results in some of the electron charge on the hydrogen being withdrawn, leaving the hydrogen slightly positively charged. If you look at the structure of the amino acids in a polypeptide chain, you will notice that this situation occurs at the regularly spaced amino groups (-NH). Meanwhile other chemical groups, most notably the equally regular carbonyl groups ($C=O$), are slightly negatively charged due to the presence of a pair of electrons. As a result, the positive amino groups are attracted to the negative carbonyl groups. And since the polypeptide chains are bristling with these charges, the chain effectively sticks to itself.

Meanwhile, Robert Corey provided the shape of the twenty puzzle pieces, having determined the structure of the individual amino acids. Together, in 1950, Pauling and Corey combined the knowledge of hydrogen bonds with the physical limitations of the amino acid structures and, piecing them together like a three-dimensional molecular jigsaw puzzle, hit upon three 'secondary structures'. The first they called an alpha-helix, which consists of a spiral-staircase-like arrangement. In an alpha-helix side chains point outward like steps (Figure 3). The structure is held together by hydrogen bonds between amino acids immediately above one another in the helix and spaced three or four amino acids apart in the primary sequence. The second is

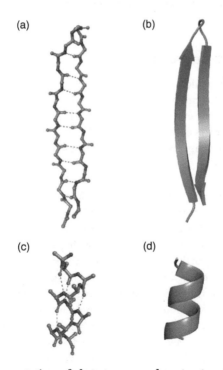

3. Representations of a beta-turn secondary structure (a) and an alpha-helix (b); a stick diagram showing bonds and atoms within the main chain, dotted lines representing hydrogen bonds between carbonyl and amide groups (c); the same stretch of protein now represented with a simpler 'cartoon' diagram (d).

the beta-strand (Figure 3): in this structure the polypeptide chain is extended. Several beta-strands sit next to one another and can zip together via hydrogen bonds to form beta-sheets. Incredibly, Pauling and Corey made these predictions without any data from an intact protein, yet when the first high resolution structures of proteins were determined, years later, sure enough there were the alpha-helices and beta-sheets. The astonishing accuracy of these predictions also means biochemists today tend to forgive (or overlook) a mistake in another of Pauling's and Corey's predictions: their third secondary structure, the gamma-helix, has never been spotted in nature.

High resolution structures

Determining the structure of proteins and so confirming Pauling and Corey's predictions required significant development of the method itself (X-ray crystallography) that Corey had used to derive the structures of the amino acids. Proteins are significantly bigger and more complicated than amino acids, and so they posed major challenges.

X-ray crystallography was conceived of in the early 20th century by a father-and-son team. William Henry Bragg and his son William Lawrence Bragg prepared a clean crystal of table salt and shone X-rays through it, which resulted in a geometric pattern of spots on their detector. Others had carried out similar experiments before, but the Braggs made a crucial intuitive leap. They realized that tucked away in the arrangement and intensities of the spots was information about salt's molecular structure. Lawrence Bragg then came up with a mathematical formula, now known as Bragg's law, that could be used to extract this information, allowing him to work out how the atoms of sodium and chlorine are arranged in a salt crystal.

Today we know that crystals consist of molecules with their atoms arranged in regular patterns. X-rays hit the electrons of these regularly arranged atoms and are scattered, going on to interact with other scattered X-rays, forming a diffraction pattern, as found when any waveforms encounter obstacles. It is this diffraction pattern that was captured on the photographic plate by the Braggs.

The link between a diffraction pattern and a structure can be difficult to comprehend. To help illustrate it let us consider a familiar and relatively simple molecule: deoxyribonucleic acid (DNA) (we'll get to its history shortly). There is an image that is very familiar to biochemists: 'Photo 51', as it has become known, is a diffraction pattern of DNA collected by Raymond Gosling, under the supervision of Rosalind Franklin, in 1952 (Figure 4). To the untrained eye it is just an esoteric spotty cross. It is difficult to see how the structure of DNA could have been inferred from it. But to Franklin and colleagues it was evident that the crystal must be a helix.

This leap from a spotty cross to a helix may seem a stretch but it is in fact quite easy to demonstrate. All you need is a laser pointer and a spring from a retractable ballpoint pen. Just shine the laser light through the spring and onto a wall about 3 metres away. You should see a cross shape that looks strikingly similar to Photo 51. The cross is the result of the way that the laser light is diffracted by the spring, and with a few simple equations you could work out the shape of the spring from the features of the cross on the wall. In an analogous fashion Photo 51 contains information that directed the building of the double helical model of DNA.

Photo 51 undoubtedly represented a ground-breaking moment in structural biology, but the information it provided was limited. It did reveal the gross helical structure of DNA but not the arrangement of the atoms within the molecule. In the case of DNA, James Watson, Francis Crick, and Maurice Wilkins could make a very educated guess at the atomic arrangement by arranging cardboard models of the DNA's chemical components until they fitted into the gross structure predicted by Franklin and

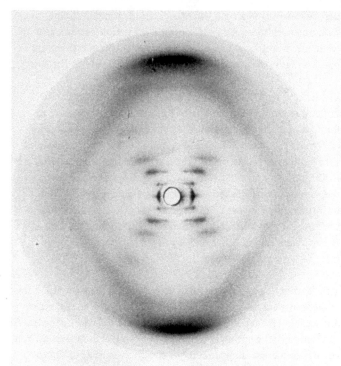

4. Photo 51, Franklin and Gosling's X-ray diffraction pattern which revealed the double helical structure of DNA.

Gosling's data. Proteins however are much more complicated beasts, and to build their detailed structures required atomic scale resolution of diffraction patterns.

As we have already seen, proteins can form crystals; in the 1920s Sumner and Northrop produced crystals of urease and pepsin to prove enzymes are proteins. Well before then, in 1839, the first protein crystals, of haemoglobin, were observed. So it was not much of an intuitive leap to expose protein crystals to X-rays in an attempt to determine their structures. The first serious attempt at this was made by Dorothy Crowfoot Hodgkin and John Bernal (himself a student of William Bragg) in 1934. They replicated Northrop's pepsin crystals and then directed their X-ray beam through them. The result was the first, wonderfully sharp, diffraction pattern of a protein.

The sharpness of the pepsin image, and those of other proteins that soon followed, caused a great deal of excitement, as it was possible to determine interatomic distances within a protein. Nevertheless, it must have been very frustrating to be faced with the spots and their intensities knowing that they encoded the location of the atoms within a protein, and yet still not having the tools to fully decipher the patterns. The crux of the problem was that there was, at the time, no way to use this information to actually locate the atoms relative to each other.

The issue was eventually overcome by replacing some atoms in a protein with heavy metals. The heavier atoms scattered the X-rays much more strongly than the lighter native atoms of the protein, thereby allowing them to be distinguished from one another. In practice this approach was extremely challenging, as there was a significant chance that the metal replacement would disrupt the structure of the protein. So, the technique required multiple replacements, and comparisons with the native proteins' diffraction patterns, to find the sites on a protein which could tolerate the imposition of a heavy atom.

As well as these technical solutions, significant computational and mathematical advances were also needed to deal with the many thousands of atoms that make up a protein and the vast number of ways that these could be arranged. Which meant that there were twenty-six years between that first sharp image of pepsin and the first high resolution structure of a protein. In 1961 the structure of myoglobin was published by John Kendrew (Figure 5), followed a few years later by David Phillip's structure of lysozyme and Max Perutz's structure of haemoglobin (a good 129 years after

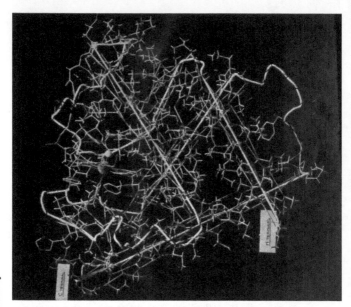

5. The first high resolution protein structure published in 1961: myoglobin.

that particular protein was first crystallized). In the intervening years Dorothy Hodgkin explored the structure of simpler molecules, including the wonder medicine of the age: penicillin. But she retained a fascination with proteins, and particularly insulin. Her thirty-five-year quest to find the structure of that particular hormone was finally realized in 1969.

The contribution that X-ray crystallography has made to structural biology cannot be underestimated. Since the Braggs' time almost thirty Nobel Prizes have been awarded (to bioscientists including Perutz, Kendrew, and Hodgkin) for discoveries directly resulting from the use of crystallographic methods and techniques, including the determination of the structures of many biological molecules such as vitamins, antibiotics, proteins, and of course DNA.

Those first high resolution structures, of myoglobin, haemoglobin, and lysozyme proteins, revealed the twists and turns of the polypeptide chain and confirmed Pauling and Corey's secondary structure. They also revealed the next level of protein complexity, the three-dimensional, tertiary structure. With these structures, finally, biochemists could examine a protein in its entirety. This allowed them to understand the molecular mechanism underpinning their workings and the chemistry of life that they controlled. Prior to this, trying to understand proteins was akin to divining the workings of a combustion engine by examining all the parts separately—now the fully assembled machine could be examined. And we'll do just that in Chapter 3.

DNA

The story of DNA is somewhat simpler than that of proteins, largely because its importance was missed for quite some time, and because it is a less complex molecule.

The 'father of genetics', Gregor Mendel, conducted his famous experiments on inheritance in 1843 with peas. Of course, at this point the molecules and mechanisms of inheritance were unknown—unequivocally linking inheritance with DNA would not happen for well over a further hundred years. The first tentative steps down that path were taken, not long after Mendel's work, by Johann Friedrich Miescher, a student in Tübingen, Germany, in 1869. He had been tasked with investigating the chemical composition of white blood cells. Unfortunately for him the best sources of these were the pus-soaked bandages of recovering surgical patients (he did later turn his attention to a much more pleasant source of cells: salmon roe). After removing proteins and lipids, he came across another substance that behaved quite differently and had a distinct chemical composition. Unlike the proteins, it was particularly rich in phosphates. And since this new material appeared to reside in the nucleus of the cells, he called it nuclein. Some twenty years later, Miescher's student Richard Altmann demonstrated that nuclein was acidic and renamed it nucleic acid.

The next great step forward came from Phoebus Levene, a prolific chemist who published over 700 papers during his career. Over a period of two decades, at the beginning of the 20th century, Levene investigated the building blocks of nucleic acids and revealed that they are composed of just three distinct chemical components: a phosphate group, a five carbon sugar (ribose), and one of four nitrogen-containing ring structures, called bases (guanine, adenine, cytosine, and thymine in the case of DNA)—the bases giving us the familiar 'letters' A, T, G, and C that define a DNA sequence. Just as importantly, Levene also worked out how the three components were linked together to form what he called nucleotides (we will look at the structure of these in Chapter 4).

In the early 20th century, biochemists were inching towards a link between DNA and inheritance. Walter Sutton and Theodor Boveri had independently identified chromosomes as the carriers of genetic material. And the great Russian biologist Nikolai Koltsov came uncannily close (in 1927) to describing the mechanism by which DNA replicates and stores information. He predicted that inheritance would occur via a 'giant hereditary molecule' consisting of 'two mirror strands that would replicate in a semi-conservative fashion using each strand as a template'. But despite his insight into the mechanisms of inheritance he was not able to go against the dominant scientific thinking of the time, so he still presumed that proteins were the agents that stored genetic information.

This presumption largely persisted until 1944 when Oswald Avery, Colin Macleod, and Maclyn McCarty carried out a meticulous study showing that the characteristics (phenotype) of bacteria could be altered by 'transforming' it with DNA and **only** DNA. The final nail in the coffin for the theory of protein-based inheritance (until the epigenetic revolution, but that's for another

Very Short Introduction) followed a few years later with the elegant work of Martha Chase and Alfred Hershey (Figure 6). At the time it was known that bacteriophages (viruses that infect bacteria) consisted of a protein shell encapsulating DNA. Chase and Hershey used radioactive isotopes of sulfur (^{35}S) to label the protein and phosphorus (^{32}P) to tag the DNA. They then infected bacteria with their labelled phage. The infected bacteria replicated more phage, indicating the genetic material had been transmitted to the bacteria. But traces of radioactivity could only be detected in bacteria that had been infected with labelled DNA, and not the labelled protein. Thus, Chase and Hershey conclusively showed that DNA was the molecule of inheritance. The work was recognized with a Nobel Prize—for Hershey. Martha Chase did not even earn a mention in Hershey's acceptance speech.

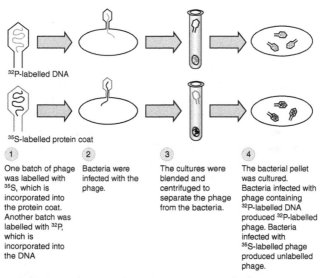

^{32}P-labelled DNA

^{35}S-labelled protein coat

(1) One batch of phage was labelled with ^{35}S, which is incorporated into the protein coat. Another batch was labelled with ^{32}P, which is incorporated into the DNA

(2) Bacteria were infected with the phage.

(3) The cultures were blended and centrifuged to separate the phage from the bacteria.

(4) The bacterial pellet was cultured. Bacteria infected with phage containing ^{32}P-labelled DNA produced ^{32}P-labelled phage. Bacteria infected with ^{35}S-labelled phage produced unlabelled phage.

6. **The Hershey-Chase experiment. Bacteria were infected with phage radiolabelled protein, or DNA. Only the ^{32}P entered the bacterial cells, which demonstrated that DNA is genetic material, and not protein.**

Despite Chase and Hershey's experiments it still was not apparent how DNA might carry genetic information. A major clue came from Erwin Chargaff's work in 1950. He found that no matter what organism DNA came from the amount of adenine equalled the amount of thymine, and the quantities of cytosine and guanine were also equivalent. This was crucially important, a year later, for James Watson and Francis Crick as they embarked on their structural determination of DNA. Chargaff's rule constrained how the famous pair could build their models (for which they used simple cardboard cut-outs to represent the bases and then hung them on wire-frame phosphate backbones), for it meant that adenines must pair up with thymine and likewise for cytosine and guanine. This immediately suggested two linked strands of DNA with the bases facing in towards each other, with the strands zipped together with hydrogen bonds. This, together with the Franklin and Gosling X-ray crystallographic data, suggested the DNA consisted of two strands coiled around one another to form a double helix.

It is worth noting that at the time there was a great deal of interest in the structure of DNA, and several other groups were racing the UK team to the solution—not least the great Linus Pauling and Robert Corey. In fact they thought they had the structure cracked and rushed out a publication in February 1953, which described a three-stranded helical DNA, with the phosphates at the centre and the bases pointing outwards. This strange structure was all the more bizarre because, in his haste, Pauling had somehow managed to overlook the very hydrogen bonds that he himself had first described.

Just a few days after Pauling's rather rushed and poorly thought through publication appeared in press, Watson and Crick were confident enough to announce their own, now iconic, structure of DNA. Famously, they did so, not in the limelight of a scientific conference or on the pages of a hallowed scientific journal, but instead to the assembled patrons having lunch at the Eagle Pub in

Cambridge. Not long after, they did use more appropriate methods to disseminate their work. In April of 1953, along with Wilkins, Gosling, and Franklin, they published a series of three back-to-back papers in the journal *Nature* which described the structure of DNA and how it was elucidated. From that double helical structure, it became immediately apparent how DNA stored genetic material. And indeed Kolstov's description of a giant molecule with mirrored strands, one serving as the template of the other, proved absolutely correct (even if Kolstov had got his macromolecules wrong).

So by the mid-20th century the structures of the two massive molecular players, protein and DNA, along with DNA's cousin, ribonucleic acid (RNA), and its myriad roles (we'll get to those later) were in place. It was becoming apparent that these were the fundamental molecular machines that marshal the chemistry within cells.

Chapter 2
Water, lipids, and carbohydrates

Before we delve further into the world of proteins and nucleic acid and get swept up in their beautiful complexity, we should take a look at some simpler but no less important molecules—molecules which form the physical structures that bound cells and provide the very medium in which the chemistry of life takes place.

Water

As we hop between branches of the tree of life we see much variation, but one molecule remains an absolute constant. Wherever water is found, on planet Earth at least, we also find life and, without it, life fails to flourish. So, it is well worth considering water's role in biochemistry and what makes it uniquely able to support life (as we know it).

Given the ubiquitous nature of water it is easy to overlook its remarkable and frankly odd properties, without which life could not exist. As you sip an iced drink on a summer's afternoon one such property is perfectly obvious: ice floats. Given how familiar we are with it, floating ice may not seem in the slightest bit odd. However, finding a solid form of a material bobbing on the surface of its liquid state is actually highly unusual. To illustrate how important this property is let us conduct a quick thought experiment. Imagine, if ice were denser than liquid water, as ice formed it would sink to the bottom of the lake or sea. Before long the body of water above would freeze solid, trapping any living things within it. Fortunately, thanks to the anomalous behaviour of water, floating ice creates an insulating layer on the surface of water, allowing the life below to survive.

Second, for a molecule of its size, water is liquid at remarkably high temperatures and over a very wide (100°C) range, which conveniently coincides with the ambient environment on most of our planet. By comparison, similar molecules such as hydrogen sulfide (H_2S) (melting point -84°C and boiling point -62°C) and hydrogen selenide (H_2Se) (liquid between -42°C and -64°C) are each liquid over just a 20°C range and at sub-zero temperatures.

These two physical characteristics are clearly key to sustaining life, but a third chemical characteristic is equally important and fundamental to biochemistry—water's structure and composition makes it an astonishingly good solvent. Each hydrogen atom in H_2O shares a pair of electrons with the oxygen atom between them. Oxygen is electronegative, so it draws some of the electrons' negative charge towards the centre of the molecule, leaving the hydrogen atoms with a slight positive charge. Chemists describe molecules with this sort of charge distribution as being polar. And the result is that these polar compounds are equally adept at interacting with both negatively and positively charged molecules. Take common table salt, sodium chloride ($NaCl$), as an example. When this is added to water it ionizes into Na^+ and Cl^-. Water forms what is known as a solvation shell around the positive sodium ion by pointing its negatively charged oxygen towards it, while it copes with the negative chloride by surrounding it with positive hydrogens. Similarly, when a molecule has a mixture of positive and negative charges on its surface, the water molecules adapt their orientation to completely envelop the molecule. We are so used to seeing salt and sugars seemingly disappear in water that it seems unremarkable. But nothing else dissolves molecules quite like this, thus making them so readily available to the biochemistry of a cell.

A second consequence of this polarity is that water can participate in the weak, but very important, hydrogen bonding. Here the partially positive hydrogens are 'donated' to the partially negatively charged 'acceptor' such as oxygens in water. Each water molecule can donate and accept two hydrogens, creating four hydrogen bonds. This creates a network of interactions that results in the high melting and boiling points of water, compared to similar molecules. And it also allows protons to hop between water and other molecules, a phenomenon that crops up time and time again during biochemical reactions.

The network of hydrogen bonds also leads to the hydrophobic effect, which is fundamental to how biological structures form. You will have seen this effect in action when making a salad dressing. Water's failing as a solvent is that it can't mix with non-polar, fatty materials. Oils, fats, and waxes do not have many (or sometimes any) charged groups to which the polar parts of water can attach. Their presence in bulk water disrupts the hydrogen bonding network, which leads the hydrophobic material to cluster together, so minimizing the area in contact with the water. This hydrophobic effect does not just drive the separation of the oil from water in your French dressing, but also results in the self-assembly of biological membranes and proteins, of which more shortly.

Lipids

Lipids are a highly diverse class of molecules, but broadly speaking they all have a significant hydrophobic component, which may take the form of hydrocarbon chains or rings, and a smaller hydrophilic (polar) group. Lipids play three main roles in biochemistry: energy storage, signalling, and structure formation.

Most lipids are based on fatty acids, formed from a 12' to 20' carbon hydrophobic tail and a carboxylic acid (-COOH) head group. When we come across terms such as saturated, unsaturated, and omega-3 oils or fats in diet plans, it is the structure of the tails of these fatty acids that is being described. Saturated fats have a hydrocarbon tail where carbons are attached to one another in a linear chain, linked by single bonds. Each carbon in the middle of the chain also covalently bonds to two hydrogens. In the case of unsaturated fats two or more of the carbons in a chain are attached by double bonds, putting a kink in the chain. Each of the carbons in a double bond can then bond to only one hydrogen. Hence these chains are not saturated with hydrogens, leading to the term unsaturated fats. Terms such as omega-3 indicate where in the chain the double bond appears. In this case the double bond is three carbon molecules from the end of the chain.

Triglycerols are lipids composed of three fatty acid chains attached to a glycerol molecule. Their primary role is energy storage. Once formed they are squirrelled away in specialized fat cells known as adipocytes. A second common type of lipid are glycerophospholipids. These consist of two fatty acids, again linked to a glycerol, but instead of the third chain a large phosphate head group (PO_4^-) is also attached to the glycerol. This creates a highly *amphipathic* molecule—meaning one end is hydrophobic and the other hydrophilic—and the mixed nature of the lipid results in something rather interesting. When placed in an aqueous environment the hydrophobic effect drives the tails together leaving the hydrophilic head groups to interact with the water. The result is the formation of sheet-like structures called bilayers (Figure 7). The hydrophobic core of these bilayers makes the whole structure impermeable to water. A bilayer can wrap around to form an enclosed bag, and such a bilayer forms an effective boundary around a cell, separating the complex chemistry within from the chaotic environment without. Of course, a cell does have to interact with its surroundings; it needs to sense what is going on around it, extract nutrients from the environment, and interact with other cells. These interactions are

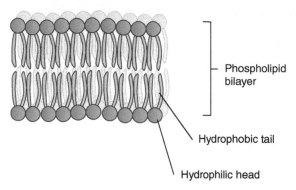

Phospholipid
bilayer

Hydrophobic tail

Hydrophilic head

7. Cartoon representation of a lipid bilayer.

mediated by proteins embedded within the lipid bilayer, where they act as pores, gates, and receptors channelling the flow of materials and information across the cell membrane. Together these membrane proteins and lipids form a two-dimensional fluid in which the components largely move freely within the plane of the membrane.

Simple-celled prokaryotic organisms such as bacteria use lipid membranes (along with complex peptidoglycan cell walls) as part of the boundary of the cell. Organisms with more complex cells, eukaryotes such as plants, fungi, and animals, further partition their cells with lipid membranes forming organelles. Each organelle has a distinct role and biochemistry, such as the cell nucleus, where DNA is housed and replicated; mitochondria, the powerhouses of cells; and in the case of plants, chloroplasts, where photosynthesis takes place.

This all sounds quite straightforward, but in actuality cell membranes are made up of a vast array of different lipids, which impart chemical and physical characteristics to the membranes. For example the kinks caused by unsaturated chains affect the fluidity of the lipid membrane. Meanwhile the head groups can also vary in size, charge, and structure, all of which can alter the function of proteins within the membrane or form discrete patches of membranes with their own specific properties.

The other major lipid component of eukaryotic cell membranes are sterols, of which cholesterol is the most familiar. These are four-ring molecules with a short chain at one end and a very small polar head group (usually just an -OH) at the other. As a result sterols are very hydrophobic and so partition into the chain region of a membrane. Their presence stiffens up a membrane. They may also be instrumental in forming nanometre-size patches, known as lipid rafts, in which particular proteins cluster, so creating distinct areas of functionality on the surface of cells. The nature of these rafts has been a matter of considerable debate. Some, with valid arguments, dispute their existence at all. But, if they exist, then rafts may serve to recruit and concentrate proteins involved in passing chemical signals through the cell, a process known as signal transduction. It is also interesting to note that rafts may be invasion points by viruses. For example, virologists have observed that in some cases removing cholesterol from a membrane (and so abolishing rafts) leaves the Human Immunodeficiency Virus (HIV) unable to infect the cell. Despite its bad press, cholesterol is a vital component of cell membranes and is a precursor for many signalling molecules such as vitamin D and the hormones testosterone and oestradiol. So removing it from your system isn't going to be a viable way to protect you from viral infections.

Carbohydrates

Carbohydrates are familiar biomolecules but still need some introduction. Frequently overlooked for the seemingly more exciting proteins and nucleic acids, carbohydrates are fundamental to biochemistry. They provide the fuel that powers cells, form the scaffolding around which so many structures are built, and frequently embellish proteins, modifying their behaviours or adding functionality.

We can classify biological carbohydrates into three main classes. The first is simple sugars, involved in energy conversion, which include single sugars, termed monosaccharides, such as glucose, fructose, and galactose. When two such sugars link together, they form the second class of carbohydrate: disaccharides, such as sucrose (formed by conjoining glucose and fructose) or lactose (made by linking glucose with galactose). Simple sugars are quick and easy for cells to metabolize, releasing the chemical energy trapped within them. However, they are difficult to store, so our bodies link sugars together to form larger polysaccharides (the third class of carbohydrate), called glycogen, while starch plays a similar role in plants. Other carbohydrates include chitin, formed from acetylglucosamine, which makes up the cell walls of fungi and the exoskeletons of insects. Keratan sulfate crops up in corneas, cartilage, and bone.

One polysaccharide in particular deserves special mention, for it is the most abundant polymer on the planet and responsible for the world's largest biological structures. Cellulose is formed from long, linear chains of glucose sub-units, each one cross-linked to neighbours via hydrogen bonds. It is from cellulose that the walls around plant cells are created, giving them their firmness. This is what makes plants such useful materials for constructing everything from furniture to clothing. Wood (along with lignin, another complex carbohydrate), hemp, and cotton are all predominantly cellulose. And in fact, a bit of biochemistry happens every time you iron a cotton shirt. The combination of heat and moisture quickly breaks the hydrogen bonds. As you apply these with a bit of pressure, all the cellulose molecules are forced to lie parallel with each other, so flattening the cloth. Then, as the shirt cools, the hydrogen bonds reform, locking in the new pressed state.

Biology is not fussy about keeping these various classes of biomolecules separate from one another. Instead it frequently links them together to form complex conjugates, mixing and matching chemical groups. We have already seen how ribose sugars are integrated into the structure of DNA. Carbohydrates also crop up as additions to proteins and lipids forming glycoproteins and glycolipids. The ABO blood group is a prime example of this. Red blood cells are covered with a variety of glycolipids and glycoproteins. Those with an O-blood group have glycoproteins bedecked with two galactose, a fucose, and n-acetylglucosamine sugars; B-blood groups have an extra galactose; and those with A-blood types have an extra n-acetylgalactosamine. Meanwhile other mixtures of carbohydrates with amino acids produce bacterial cell walls; peptidoglycans are polysaccharides linked with short stretches of amino acids. Interestingly, a bacterial protein that creates this cross-link is the target for the antibiotic penicillin; it stops bacterial cell walls from forming.

Energy currency and electron shuttles

Many biochemical processes are energetically unfavourable: they are the metabolic equivalent of throwing a stone uphill, and so they require quick energetic kicks to get them moving. Most of the energy to power these processes ultimately comes from the Sun. The energy is then trapped by plants through photosynthesis before being stored as high energy electrons within the bonds between the carbons of carbohydrates and fats. However, releasing the energy locked away in those stores is a slow process. You can think of the carbohydrates and fats as energy that is 'banked' in a savings account, where you have to make a special effort to get at it. Cells need a much more 'liquid' source of energy to power the cellular pumps, motors, and enzymes and other components of the never resting cellular machinery. In effect they need a universal and easily exchangeable energy currency. And, across the living world, the main molecule that fulfils this role is adenosine triphosphate (ATP).

ATP consists of adenine (which we have already encountered as one of the four bases in DNA), attached to a ribose sugar and a row of three phosphate groups. The last phosphate group is prone to undergoing enzyme-catalysed hydrolysis (i.e. enzymes can remove it) to produce a free phosphate, water, and adenosine diphosphate (ADP). In the process energy is released, which can be used to power the machinations of the cell. At the core of the metabolic process are pathways and cycles dedicated to breaking into the energy stores of carbohydrates and fats and using them to recharge the stocks of ATP (more on this in Chapter 5). The upshot is that every ATP molecule might undergo this recycling process some 2,000 to 3,000 times a day.

ATP has a set of sister molecules that play one more essential role in biochemistry. ATP (adenosine triphosphate), GTP (guanine triphosphate), CTP (cytosine triphosphate), and UTP (uracil triphosphate) are the nucleotides used to construct RNA. Meanwhile, deoxyadenosine triphosphate (dATP), deoxyguanine triphosphate (dGTP), deoxycytosine triphosphate (dCTP), and deoxythymine triphosphate (dTTP) are used to construct DNA (Figure 8).

It is sometimes said that chemistry is just the movement of electrons, because electrons are the medium of chemical bonds; as they move, bonds are formed and broken. So constructing something as complicated and dynamic as the chemistry of life needs a lot of shuttling of electrons. Another group of adenine-based molecules are responsible for most of the relaying of electrons from one reaction site to another, and so plays pivotal roles in biochemical processes. These electron shuttles are nicotinamide adenine dinucleotide (NAD), flavin adenine dinucleotide (FAD), and nicotinamide adenine dinucleotide phosphate (NADP). These all exist in oxidized forms (NAD$^+$, FAD,

8. (a) Nicotinamide adenine dinucleotide in its oxidized form; (b) cholesterol; (c) a commonly occurring unsaturated phospholipid; (d) adenosine triphosphate; (e) the disaccharide sucrose, composed of glucose and fructose rings.

and NADP⁺) which are electron acceptors, and reduced forms NADH, $FADH_2$, and NADPH able to donate electrons, e.g. $NAD + H^+ + 2e^- \rightleftharpoons NADH$. Much like with ATP, this trio of molecules are stuck in a recycling process as they shuttle electrons from one place to another and so are continuously undergoing conversions between their oxidized and reduced forms.

Chapter 3
Proteins: nature's nano-machines

Proteins are the dominant form of cellular machinery. Every biological process is regulated, performed, or governed by proteins. Proteins control the metabolic flux, provide mechanical support, form the intracellular highways along which materials are transported, manage genomes, and act as the gateways through membranes. This multitude of processes needed to keep cells functioning are managed in the organism or cell by a massive cohort of proteins, together known as the proteome. The size of a proteome will vary with the complexity of the organism: An individual mammalian cell may need over 15,000 proteins to function; a whole human requires upwards of 90,000 individual proteins. A bacterium can make do with 3,000, while a virus may need just a few dozen.

As we have seen already, proteins have hierarchical structures starting with a primary sequence of amino acids linked together into a polypeptide chain. This folds up into secondary structures consisting of alpha-helices and beta-strands linked together to form beta-sheets, as well as less defined turns that connect the other two structures. In turn the secondary structural elements come together to form an overall three-dimensional shape known as the tertiary structure. The arrangement of multiple protein chains to form complexes is the quaternary structure. The variety of structures that proteins adopt are beautiful, many, and varied.

Since 1958, when the structure of myoglobin was revealed, the numbers of proteins whose structures have been established has exploded to well over hundreds of thousands. And they can all be perused via a public-domain archive—Protein Data Bank (PDB; at rcsb.org).

Structural motifs

If you trawl through the vast PDB (and I strongly recommend you do, it really is a fascinating corner of the internet) you may notice patterns in protein folds emerging. There are, it seems, a limited number of ways by which secondary structures interconnect and these are classified into a small number of motifs (see examples in Figure 9). One of the most common motifs is the beta-hairpin consisting of anti-parallel beta-strands connected by a turn made up of a few amino acids. Similarly, two helices may be connected by a turn in a helix-turn-helix motif. Other motifs are more complicated, such as a Greek key arrangement of four beta-strands, or mixtures of secondary structural elements, such as in the zinc finger. This is a particularly common fold in proteins that interact with DNA or RNA, and it positions two beta-strands and an alpha-helix together, in such a way that two histidine and two cysteines bind a zinc ion.

The complexity of protein structures builds further, with motifs joining to create larger arrangements. Yet even at this level structural themes emerge, with particular conformations being conserved and used as scaffolding on which to hang protein-specific details.

For example, the helix-turn-helix is often seen in the descriptively named four-helix bundle. This common and simple arrangement crops up time and again. Sometimes they appear just as isolated bundles, as in the human growth hormone. Alternatively, precisely twenty-four individual bundles coalesce creating ferritin, a large ball structure that transports iron. Or they may be

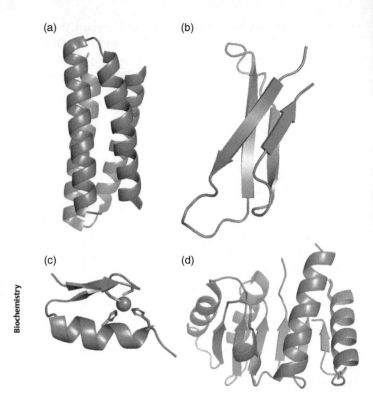

9. Common protein structural motifs: (a) two helix-turn-helix motifs forming a four-helix bundle; (b) four beta-strands creating a Greek key; (c) a zinc finger with the zinc ion held in place by histidine and cysteine side chains; (d) the more complicated Rossman fold.

just one element of a much larger protein as in the Lac repressor, a DNA-binding protein consisting of multiple different folding domains.

Meanwhile, multiple Greek keys and beta-hairpins are stitched together into extended sheets that eventually curl in on themselves to create beta-barrels. These are particularly common in bacterial membranes where they act as pores (the class of protein is known as a porin) through which molecules diffuse. Porins have evolved to be specific for a particular molecule allowing cells to carefully control the flow of chemicals across their membranes.

Then there are more complicated arrangements such as the Rossman fold, consisting of two sets of helical pairs sandwiching a six-stranded beta-sheet. The Rossman fold is frequently seen in nucleotide binding proteins and dehydrogenase proteins.

Cofactors and post-translational modifications

The twenty amino acids that make up the bulk of proteins produce the vast array of protein structures, but they do not provide quite enough chemical variety to get all the biochemistry of a cell done. So frequently proteins need something extra, in the form of additional chemical groups. Sometimes these are added shortly after proteins are manufactured (translation), and such significant alterations to proteins are known as post-translational modifications.

Glycosylation is the most common such modification. Approximately half of all proteins are modified by addition of sugars to their surface. In some cases these can add another 50 per cent to the mass of the protein in question. The roles of glycosylation are manifold, ranging from regulating protein binding to receptors, to increasing thermal stability and longevity of the protein. Similarly many proteins are regulated by the addition or removal of phosphate groups, which account for the phosphorus seen in Mulder's early elemental analysis of proteins.

Additions to proteins that are essential for their function are known as cofactors. Metal ions are particularly common cofactors. The most familiar example of this is in haemoglobin, where iron plays a vital role as a carrier of oxygen (some invertebrates use copper in place of iron, giving their blood a blue hue). Many enzymes that interact with nucleic acids require zinc (as in the zinc finger proteins mentioned above), and magnesium is needed by enzymes involved in fermentation. It is for this reason that small amounts of these metals, along with manganese, copper, potassium (although this is needed in much larger quantities because of its role in nerve signalling), nickel, selenium, and molybdenum are necessary in our diets.

Other, more complicated cofactors, often derived from vitamins, are known as coenzymes. These may be covalently attached (such as coenzyme A, derived from vitamin B, which crops up in a number of metabolic pathways ranging from fatty acid synthesis to energy production) or loosely associated with a protein such as NADH. Time and time again, NADH features throughout metabolic processes, so we will be meeting it again later. Another reoccurring class of coenzymes are the porphyrins, typically used to hold a metal ion in place (Figure 10). These are synthesized from amino acids and other simple molecules, and form the haem of haemoglobin, myoglobin, and catalase as well as the photosynthetic pigment chlorophyll.

Protein function

Given the central role that proteins play in biochemistry we will look in detail at how they interact with one another, and with other biochemicals. But first let us consider how the structure and function of proteins are linked. To illustrate the point, let us look at the protein catalase. It is a good example not least because it is so easy to see the enzyme at work. If a teaspoon of dried yeast is added to a cup of hydrogen peroxide (H_2O_2), you will immediately see bubbles forming. Those bubbles are oxygen and they are the result of the catalase in the yeast catalysing the decomposition of the hydrogen peroxide into water and oxygen. Catalase is present in every cell of every organism, and its job is to protect cells from the harmful, caustic hydrogen peroxide created as a by-product of numerous other biochemical reactions.

10. Two examples of porphyrins: (a) a haem group found in haemoglobin, myoglobin, and catalase; (b) chlorophyll A.

Before we look at catalase in more detail, it is worth spending a little time considering enzymes in general. Like all catalysts, enzymes speed up reactions, but they do not result in more product being formed from it. By analogy, imagine a reservoir, formed by a dam, high on a mountainside. Water flows through an open gate in the dam, and winds its way downhill. If more dam gates are opened the rate that the water leaves the reservoir increases, but the gates have no effect on the final state. Eventually all the water will find its way to the valley floor, it's just that opening further gates allowed this ground state to be reached much more quickly. Similarly, enzymes have no effect on the possible end point, they merely speed up the process towards reaching that end point. That's not to understate the power of enzymes—sometimes the increase in reaction rate is formidable. For example, the enzyme orotidine 5'-monophosphate (OMP) decarboxylase demonstrates astonishing catalytic efficiencies, by taking a reaction that normally has a half-life of seventy-eight million years, and reducing that time to just eighteen milliseconds. (Enzymes are generally named for the reaction they catalyse followed by the suffix '-ase'. For example, an enzyme that synthesizes ATP would be an ATP synthase, and DNA polymerase creates polymers of DNA.)

Catalase is also a wonderful example of the efficiency that proteins are capable of. A single catalase enzyme can decompose millions of hydrogen peroxide molecules every second. In fact the reaction rate is thought to be limited only by the speed that the water can exit the enzyme, so catalase could not physically work any faster. To achieve this incredible feat requires a large and complex structure consisting of four identical polypeptide chains, each over 500 amino acids long. The chains fold into four separate domains, a core beta-barrel surrounded by three others formed from helices and less structured loops. And nestled within the beta-barrel in each chain are four haem groups. This whole elaborate structure is some 7,000 times bigger that each H_2O_2 molecule. But it is all necessary to form a channel that guides the hydrogen peroxide to the heart of the enzyme where a pocket that perfectly matches the shape of two hydrogen peroxide molecules awaits. On the edge of that pocket is poised a histidine side chain, and a haem group tethered to a tyrosine. Each of these is held in precisely the location needed to manipulate hydrogen peroxide. And so the stage, or 'active site', is set for a molecular dance (Figure 11).

The oxygen cannot simply be ripped out of the hydrogen peroxide, instead the molecule needs to be disassembled, and the various pieces briefly held in place (in what is referred to as a transition state) before being reassembled into molecular oxygen and water. First, to access one of the oxygen molecules in the H_2O_2 the enzyme removes and holds on to one of the hydrogens. The histidine performs this role: the nitrogen in the histidine side chain is slightly negatively charged so it will attract a positive proton (hydrogen ion). This leaves the oxygen available to bind to the haem group, which leaves behind a negatively charged hydroxide (OH^-) ion. This negative charge draws the hydrogen back from the histidine, creating a water molecule. This newly formed water then swaps places with a second hydrogen peroxide. The first step is repeated with a proton briefly leaving the H_2O_2 for the histidine. This time the available oxygen binds to the first oxygen that had been deposited on the haem group, and together they form a second molecular oxygen and another OH^- ion. Finally, the hydrogen on the histidine joins the OH^- to produce another water molecule.

The intricacy and precision of the catalase reaction scheme is an excellent example of the critical link between a protein's structure and its function. An analogy that is frequently used to describe this relationship is the lock and key model (first proposed by Emil Fischer in 1894), with the enzyme presenting a rigid cavity that perfectly matches the shape of a substrate molecule. It is a nice analogy in many ways but it fails to take into account the flexibility of proteins, which is why Daniel Koshland's induced-fit hypothesis is generally preferred. In this model an enzyme and its

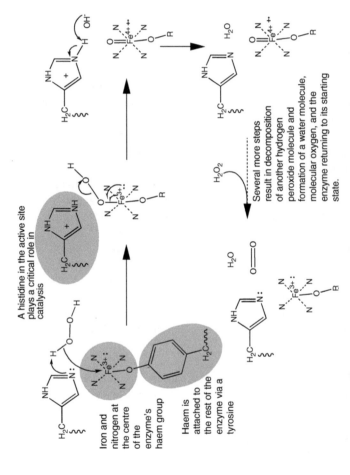

11. Catalytic mechanism of catalase (the curved arrows represent movement of electrons).

substrate are more akin to a glove and a hand, each flexing to accommodate the structural changes in the other, until they fit together, like a glove. This flexible link between structure and function will become increasingly more apparent throughout the book.

Enzyme kinetics

In Chapter 1, we saw how X-ray crystallography gave biochemists their first glimpse of the atomic structures of proteins. This allowed them to begin to understand how all the amino acids, cofactors, and ligands (substances that bind to a biomolecule) come together to drive biochemical processes. However, most structural techniques, and particularly crystallography, provide static pictures of proteins. To fully understand how dynamic systems such as enzymes work they also need to be studied in action. A key example of this is the field of enzyme kinetics.

Since the mid-19th century, chemists had been studying reaction rates. They frequently found the speed of a reaction was proportional to the concentration of the reactants. However, the same did not seem to apply to biological systems, for example yeast fermentation rates appeared to be independent of the amount of sucrose present. Then, in 1913, Leonor Michaelis and Maud Leonora Menten noticed that this was not strictly true. At low concentrations of substrates, the reaction rate was dependent on the amount of substrate present, and the reaction rate then plateaued at high concentrations.

To explain this phenomenon, which was unlike the behaviour of other chemical reactions, Michaelis and Menten proposed that an enzyme (E) and substrate (S) form an intermediate enzyme–substrate complex (ES). The intermediate then breaks down, releasing products (P) and free enzyme:

$$E + S \rightleftharpoons ES \rightarrow E + P$$

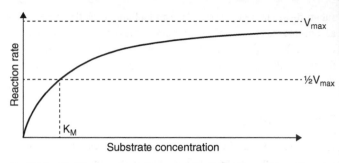

12. Michaelis-Menten saturation curve depicting the relationship between an enzyme's reaction rate and substrate concentration.

The consequences of this model are that at low concentrations of substrate the reaction rate is limited by the formation of the ES complex, whilst at higher concentrations ES forms rapidly and the rate limiting step is dictated by its breakdown. This is easily visualized in what is known as a Michaelis-Menten saturation curve, which clearly shows how an enzyme's reaction rate is dependent on substrate concentration (Figure 12). It is also important to note that Michaelis and Menten's reaction scheme is frequently a vast simplification. Many enzyme catalysed reactions have multiple ES and enzyme product complexes. These will make the reaction much more complicated than the simple description given here. Nevertheless, an understanding of Michaelis and Menten's work and its implications does serve to underpin more complex descriptions of enzyme kinetics.

The curve yields several parameters that describe an enzyme's behaviour. The maximum reaction rate (V_{max}) occurs when the enzyme active sites are saturated with substrate. From V_{max} two other parameters can be derived. The Michaelis-Menten constant (K_m) is the substrate concentration at $\frac{1}{2} V_{max}$, and provides a measure of an enzyme's affinity for a substrate; a low K_m means the V_{max} is reached at low concentrations of substrate and hence the enzyme has a high affinity for that substrate, and vice versa.

The rate at which ES degrades into product and free enzyme is given by the catalytic rate constant (k_{cat}). This is determined by dividing V_{max} by the enzyme (or more accurately the active site) concentration. k_{cat} provides a measure of how quickly an enzyme can cycle through its reaction. The last particularly important parameter is the efficiency of the enzyme, which is given by k_{cat}/K_m.

The usefulness of these kinetic parameters is best illustrated by comparing a few enzymes (see Table 1).

OMP decarboxylase has an extremely low K_m, indicating an astonishingly high affinity for its substrate. However, its ES complex exists for a relatively long time, which moderates its overall efficiency (k_{cat}/K_m). Meanwhile, catalase has a much lower affinity for its substrate but is much quicker at releasing the products, so its overall efficiency is better than that of OMP decarboxylase.

An enzyme's kinetic parameters actually describe the relationship between the enzyme and a particular substrate. Chymotrypsin illustrates this very well. It is a protease (an enzyme that cuts up other proteins) with a preference for cleaving the peptide bond immediately preceding a large, aromatic amino acid side chain (such as tyrosine). However, it is capable of cleaving those between other amino acids. This specificity manifests in the parameters as a markedly higher affinity (K_m), faster rate

Table 1. Kinetic parameters for three enzyme and varying substrates, shown in brackets

Enzyme	K_m (M)	k_{cat} (s^{-1})	k_{cat}/K_m (S^{-1} M^{-1})
OMP decarboxylase	7.0×10^{-7}	39	5.6×10^{7}
Catalase	1.1×10^{-3}	1.0×10^{6}	9.1×10^{8}
Chymotrypsin (valine)	8.8×10^{-2}	0.17	1.9
Chymotrypsin (tyrosine)	6.7×10^{-4}	190	2.9×10^{5}

constants (k_{cat}), and so a significantly better catalytic efficiency for tyrosine, its preferred substrate.

The case of chymotrypsin and its preference for particular substrates illustrates how understanding the kinetics of a reaction highlights aspects of the enzyme's reactions that would not be immediately apparent from structural studies. Similarly, enzyme kinetics comes into its own when trying to elucidate the way that enzymes and other proteins are regulated. An enzyme's activity is frequently modulated through interactions with other proteins, small inhibitory molecules, or modifications. All of these will have a discernible impact on the enzyme kinetics. For example, a molecule may bind to the active site of an enzyme and so exclude the substrate. This competitive inhibition, as it is known, manifests as a decrease in the enzyme's affinity for the substrate, and so K_m increases, whilst V_{max} is unaffected. Other inhibitors may bind elsewhere and cause a change in the conformation of the enzyme. This non-competitive inhibition leaves K_m unaffected (as the active site is still fully accessible) but impairs the functionality of the enzyme and so reduces V_{max}.

The protein folding problem

In 1958, shortly before he received the Nobel Prize in Chemistry for his work on the structure of protein, John Kendrew remarked that 'The arrangement [of proteins] seems almost totally lacking in the kind of regularities which one instinctively anticipates, and it is more complicated than it has been predicated by any theory of protein structure.' This sentence summarized what has become known as the protein folding problem—in short, how does a protein, which starts life as a long, extended polypeptide chain, fold up into a well-defined three-dimensional shape in order to function? Is there a cellular assembly manual and machinery that manipulates nascent polypeptide chains and guides them to the correct fold, or can proteins fold themselves?

The latter question was answered in an experiment devised by Christian Anfinsen in 1961. He took an enzyme called ribonuclease and measured its ability to cut up RNA. He then unfolded the protein using urea and reducing agents to break the disulfide bonds between the protein's eight cysteine residues. In so doing the protein lost its shape and with it its ability to digest RNA. Next he removed the urea and allowed the disulfides to reform. He then observed, in a matter of minutes, a recovery in the enzyme's capacity to digest the RNA. This experiment may sound simple, trivial even, but its importance cannot be overstated, because it demonstrated that proteins can fold themselves, without any external assistance. And this suggests that all the information needed for a protein to fold is contained in the protein's primary sequence. The implications are important, meaning that, in turn, a protein's structure is determined by the information contained within a gene. Therefore, one could, theoretically, predict a protein's structure from a DNA sequence. This premise became known as the Anfinsen dogma.

Unfortunately, understanding that proteins can fold by themselves is a quite different thing from understanding how proteins fold, which is an astonishingly complicated problem—as illustrated by a thought experiment devised by Cyrus Levinthal in 1969. The Levinthal paradox as it has become known, is as follows:

Imagine a small protein made up of just 101 amino acids. Now let us assume that each amino acid can adopt just three orientations relative to its neighbour. In that case our model protein would be able to adopt 3^{100} or 5×10^{47} conformations. If a protein could search through all these conformations at the staggeringly fast rate of one every 100 femtoseconds (which they can't) then it would still take 10^{27} years to check through every possible shape the protein could adopt. Given that this timescale is many orders of magnitude longer than the age of the Universe and that in reality most proteins can fold in milliseconds (Anfinsen's ribonuclease is actually a rather slow folder), there must be some way in which proteins are directed down an efficient folding pathway. And if we understand that then we may find the holy grail of structural biochemistry—the ability to predict the three-dimensional structure of a protein from just the DNA sequence of the gene that codes for it.

A huge amount of experimental effort (including some of my own) has gone into attempts to map protein folding pathways. Scientists have altered amino acids, created artificial proteins, manipulated folding environments, and much more besides. Then they have observed how all this affects the folding speeds, protein stabilities, and final protein structures, using a dizzying array of techniques. As a result, we now know a huge amount about the forces involved, how the hydrophobic amino acids quickly clump together forming a core, whilst the charged amino acids sit on the surface interacting with the watery solvent. This has led to a concept sometimes called the nucleation-condensation model, whereby the hydrophobic core forms rapidly, dragging secondary structural elements into the correct orientations relative to each other.

Emerging from this are models of protein folding behaviour based around the idea of a 'folding landscape' whereby the trajectory that an unfolded protein takes towards its folded state is analogous to a ball rolling down the side of a mountain. Each spot on that landscape represents one of the myriad possible conformations described in the Levinthal paradox, the folded state being the spot at the lowest point in the valley. Whilst this is a helpful way of describing the problem, we still have a limited understanding of what a real protein folding landscape looks like and how proteins are directed down exactly the right path (or paths) to the folded state.

The protein folding problem is likely to be solved *in silico* with a vast amount of computational power. To that end one of the most significant computational protein folding projects is something to which anyone's computer can contribute. Since 2000, the fold-at-home project has distributed protein folding problems to owners of idle computers around the globe. At the time of writing many hundreds of thousands of processes are contributing to the projects, clocking a processing power of over 1.5 exoflops. By comparison the world's most powerful supercomputer (IBM's Summit) has a tenth of the computing power, peaking at about 143 petaflops.

As in every other area of technology, algorithms and hardware dedicated to the protein folding problem are developing rapidly, but there are, broadly speaking, two approaches. The first uses template-based models, where the target protein sequence is first compared to others of a known structure and that information is then used to constrain the structural model of the target. This method produces the most accurate results, but does not really provide any insight into folding landscapes. Instead, the second, *ab initio* approach aims to create a complete solution to the protein folding problem built from scratch and based on what we know about the physics and chemistry within a protein folding environment.

Chaperone proteins

Before we finish with protein folding, it is worth mentioning a class of proteins that are vital for cellular viability. These proteins aid the folding of other proteins, and so are known as chaperone proteins. On the face of it their existence seems to contradict the premise of the Anfinsen dogma. However, we need to remember that the environment in the cell is far from that of the test tube. Cells are crowded places, and the proteins within them are subjected to changes in temperature and chemical stresses, all of which can affect the way a protein folds. As a result, a cell can rapidly accumulate misfolded proteins. Chaperones help to clear up the mess.

A particularly well-studied chaperone is a huge protein complex made up of fourteen individual polypeptide chains, known as GroEL. This barrel-like structure can completely engulf a misfolded protein. When it does so, a second complex called GroES caps off the barrel. Together they utilize the chemical energy stored in ATP to mechanically unfold the protein within and then allow it to reattempt folding. The GroES cap then detaches and the newly folded protein is released back into the cellular environment where it can get on with its job. Chaperones do not actually contradict the Anfinsen dogma as they do not provide any additional information, but instead they are like guides that direct a protein back to the start of the path down the valley, and along the correct pathway through their energy landscapes.

Intrinsically disordered proteins

The analogy of proteins as nano-machines, or nature's robots, is compelling. Just like robots, many proteins are engineered (by evolution) to repeat a set task over and over again. The analogy also implies that proteins all have a defined structure, inextricably linked to their function. Indeed, this was the dominant dogma throughout the 20th century, cemented by the thousands of clean structures swelling the protein databank. And so, the paradigm that amino acid sequence defines the three-dimensional structure, which in turn leads to protein function, became firmly entrenched in protein biochemists' psyche. In no small part this was due to the dominant technique used to determine protein structures—X-ray crystallography.

As prolific as the use of the technique has been, it has a serious limitation—structures can only be determined if one is able to grow high quality crystals of the molecule of interest. And that is far from a trivial task; whole careers have been consumed with attempts to crystallize particularly recalcitrant proteins. The problem stems from the fact that crystals can only be grown when molecules pack together in ordered arrays. So, something with a well-defined structure is likely to readily crystallize, whilst something with an ill-defined, disordered structure is unlikely to do so. By way of analogy, some proteins are like uncooked spaghetti packed in a jar, where all the individual pieces line up rather nicely with one another; other proteins are more like cooked spaghetti poured into the same jar, forming a tangled mess with no discernible order.

The problem is that X-ray crystallography is biased towards the ordered proteins, or proteins that can be forced into an ordered state. As a consequence, the PDB under-represents what have become known as intrinsically disordered proteins. Not only that, but it is now becoming evident that by shoehorning proteins into crystals we may have overlooked the intrinsic disorder rife within biochemistry, with well over 50 per cent of proteins containing significant regions of disorder. Many proteins, it seems, are 'on the edge of chaos', flitting between ordered and disordered state as needs must. It now transpires that intrinsically disordered proteins play a multitude of roles, such as acting as molecular mortar holding together multi-protein complexes, or in some cases adopting multiple conformations and functions dependent on their binding partners. Our understanding of these overlooked and enigmatic proteins is still very much in its infancy, but what is apparent is that there is much more plasticity of protein structure and function than previously thought.

Chapter 4
Nucleic acids: life's blueprints

The central dogma of molecular biology

In 1958 Francis Crick described the central dogma of molecular biology.

> This states that once 'information' has passed into protein it cannot get out again. In more detail, the transfer of information from nucleic acid to nucleic acid, or from nucleic acid to protein may be possible, but transfer from protein to protein, or from protein to nucleic acid is impossible. Information means here the precise determination of sequence, either of bases in the nucleic acid or of amino acid residues in the protein.

Or to put it another way, a DNA sequence determines that of the amino acids in a protein, but not the other way around. With minor tweaks, most notably toning down 'impossible' to something less restrictive, the central dogma still holds.

Fundamentally the central dogma consists of: (1) DNA replication—the straightforward copying of information (using Crick's definition); (2) transcription—the process of creating RNA from a DNA template; and (3) translation—creation of a protein sequence from an RNA template. The central dogma is actually a little more complicated than this, as RNA can also act as a template for DNA and RNA. But for the stake of brevity I'll stick with these three steps. In the rest of this chapter we will take a closer look at the nucleic acid molecule's form and function and how these underpin the three trunk data flows of the central dogma.

Nucleotides, strands, and base pairs

The double helical structure of DNA makes it, arguably, the most recognizable of molecules. But the images that we so frequently see in print and on screen rarely reveal the chemical building blocks of DNA nor how they come together to form the double helix. Even more rarely does DNA's chemical cousin, RNA, get a mention, despite it being a much more versatile molecule, and having many more roles in biochemistry.

As we saw in Chapter 1, nucleic acids are constructed from nucleotides, which in turn have three chemical components: a five carbon ribose sugar, on which hangs a base (guanine, cytosine, adenine, thymine, or uracil) and a phosphate group. The base is always attached to the first carbon of the sugar (known as the 1' carbon) whilst the phosphate links to the fifth (5') carbon (Figure 13).

The difference between RNA and DNA comes down to subtle chemical alterations in the base and the sugar of the nucleotides. RNA has a hydroxyl group (-OH) at the ribose sugar 2' carbon, whilst in DNA the 2' carbon has a hydrogen atom attached. Meanwhile uracil (used by RNA) lacks a methyl group where thymine (in DNA) has one.

Strands of nucleic acids are formed by covalently linking the 3' carbon to the phosphate group of the preceding nucleotide. This allows us to define some directionality in the chain, the 3' end starts with a ribose sugar and the 5' end terminates in a phosphate group.

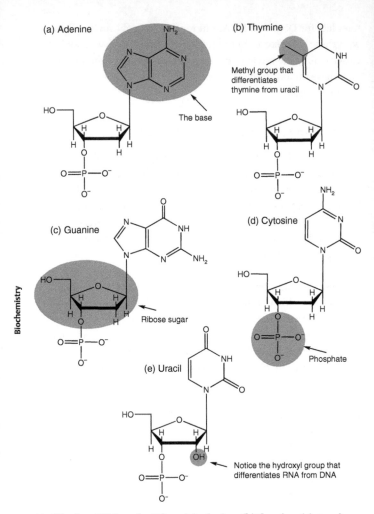

13. The four DNA nucleotides—(a) adenine, (b) thymine, (c) guanine, (d) cytosine; and one example of an RNA nucleotide—(e) uracil.

Each DNA double helix is constructed from two individual anti-parallel (meaning the 3' end of one molecule lines up with the 5' end of the other molecule) strands. These are zipped together by non-covalent hydrogen bonds formed between bases. Base pairs in DNA consist of guanine with cytosine and adenine with thymine. RNA is often single-stranded, but when it does form double strands adenine pairs with uracil.

It is this consistent G-C, A-T base pairing that allows DNA sequences to be so easily and faithfully copied during cell division. When a new copy of DNA is needed the double helix is unwound and strands separated. Each strand then acts as template from which a daughter strand is synthesized. This process of building a new DNA strand from a template is carried out by a class of enzymes called DNA polymerases. Since G always pairs with C and A with T, where there is a G on the parent strand a DNA polymerase places a C on the daughter strand. And likewise for A and T. On this level the process sounds simple enough, however it actually involves at least another twenty proteins working in conjunction with each other. This astonishing complex is known as the replisome and will be the subject of Chapter 6.

Nucleic acid secondary structures

We have already taken a look at primary, secondary, and tertiary elements that make up the structural hierarchy of proteins, and similar categories exist for nucleic acids. In the case of nucleic acids, the primary structure is defined by the nucleotide sequence instead of amino acids.

In proteins, the secondary structure is defined by hydrogen bonding between amino and carbonyl groups, which gives rise to the alpha-helices and beta-sheets. In the case of nucleic acids it is the base pairing interactions that define the secondary structure. The simple aesthetics of the DNA double helix is the most obvious example of nucleic acid secondary structure. In fact, there are

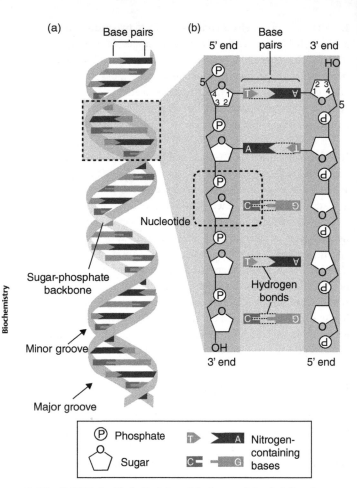

14. The B-DNA double helix (a), and the base pairing holding the two strands together (b).

three biologically relevant helical structures that DNA can adopt. The most common, and the structure described by Watson, Crick, and Wilkins, is B-DNA (Figure 14). This form is a right-handed helix (the chain twists in a clockwise direction as you look down the shaft of the helix), one turn takes place every ten nucleotides, and the double helix has two noticeably different-sized grooves: the major groove and the minor groove. A-DNA was discovered by Rosalind Franklin and Raymond Gosling in their crystals. This form of DNA frequently forms as DNA is dehydrated, and it turns out that it is adopted in desiccated bacterial spores. A-DNA is also right-handed, but in comparison to the B-form it is wider and looks more like a compressed spring. Meanwhile the Z-form is an elongated left-handed helix and unlike the A and B forms there is barely a difference between the major and minor grooves. The A and B forms crop up in normal biological functions, whilst Z-DNA only appears in some diseased states.

In contrast, RNA is often single-stranded, although molecules frequently contain regions with complementary sequences, giving rise to intrastrand interactions. These result in a range of secondary structures including double helices, which in the case of RNA tend to resemble A-DNA. These commonly crop up in hairpins, which feature a short base paired helical stem topped by a loop of unpaired bases. Further secondary structural complexity arises from interactions between hairpins, when portions of loop base pair to one another creating pseudoknots.

RNA world

At the heart of the central dogma there is a chicken and egg situation. Life needs molecules that, together, have the capacity to produce more molecules like themselves. Proteins can do half of the job. They are excellent 'manufacturers' with the ability to form the various structures and associated functions needed to catalyse a vast array of chemical reactions. For example, they can stitch together nucleotide bases into stretches of DNA and RNA. But there is no known mechanism by which proteins can make copies of themselves. Meanwhile, DNA makes an excellent data storage molecule, with a structure that lends itself to a straightforward replication mechanism. But DNA cannot catalyse reactions, so without the proteins a DNA sequence will never be copied and inheritance would be impossible. In short, gene replication requires proteins, and proteins cannot be manufactured without the information in a gene. Given this need for both DNA and proteins to work together, how could this synergistic system have begun?

There are several theories, including co-evolution of DNA and proteins or a 'proteins first' hypothesis, where small proteins did replicate themselves using a mechanism now lost in the mist of evolutionary time. However, the dominant theory lies with a molecular all-rounder, a theory known as the RNA world. In this scenario, before the advent of today's protein and DNA filled cells there was a form of life that relied on RNA. This works because RNA exhibits some of the properties of both proteins and DNA. It can act as a store of genetic information, and some viruses still use it as such. Moreover, the hairpins, pseudoknots, and other secondary structural elements can fold to create complex structures capable of catalysing chemical reactions.

The remains of this RNA world can still be seen lurking within biochemical corners of the cell. Some of it was revealed in the 1980s when Thomas Cech was studying the mechanism underpinning the splicing of a section of RNA. Try as he might he could not find the protein responsible for the reaction, until finally he was led to conclude that the RNA could splice itself. Cech later coined the term ribozymes (a portmanteau of **ribo**nucleic acid en**zymes**) to describe this catalytically active RNA.

Cech's and others' similar discoveries lent weight to the RNA world hypothesis, but a demonstration was still needed to show that RNA could indeed replicate itself in order to crack the chicken and egg paradox. As yet no one has seen evidence for this in natural systems, which may not be surprising given that it would have been made redundant by the much more efficient protein-based systems. However, scientists have created artificial ribozymes capable of replicating themselves, thus demonstrating that a scenario of living systems based around self-replicating RNA is at least biochemically feasible.

Transcription and gene structures

In common parlance, chromosomes, genes, and DNA are almost synonymous, but in actuality chromosomes are the structures on which genes are stored, and genes are defined as a stretch of DNA that codes for another biologically functional molecule—that is, proteins and RNA. In addition to this coding region, a gene will have a series of sequences that serve to regulate how the gene is transcribed.

When genetic information is needed, copies of the genes are made and ferried out to other parts of the cell. Messenger RNA (mRNA) is the material used to make these copies and it is manufactured in much the same way that a daughter strand of DNA is created from a template. The difference is that in this case the enzymes responsible for matching up bases (cytosine with guanine and adenine with uracil) are RNA (instead of DNA) polymerases.

The DNA strand that will be transcribed is known (slightly counterintuitively) as the non-coding strand, because by acting as a template the resulting mRNA will have a complementary coding sequence. Just as in any manufacturing process, signals are needed to indicate when to start production and how much of a product is needed. So around, and amongst, the coding section of the gene there are a range of other sequences many of which play regulatory roles. Within the gene these coding regions are prefixed with 'promotor' sequences doing exactly that.

Prokaryotes generally have two promotors located about ten and thirty-five nucleotides upstream of the transcription start point. Comparison of hundreds of genes reveals a consensus sequence of 5'-TTGACA-3' and 5'-TATAAT-3' for the -35 and -10 boxes (as they are imaginatively named) respectively. Proteins known as transcription factors bind to these promotor sites and then recruit RNA polymerases to the region. The rate of transcription of genes is regulated through variations in the promotor sequences. The more deviation there is from the consensus the lower the affinity it has for the transcription factor and so the gene gets transcribed less frequently. Conversely proteins that are needed in high numbers will have a promotor sequence closely matching the consensus.

Terminating transcription is equally as important as initiating it. Without some control RNA polymerases will charge along a DNA sequence like a runaway train, churning out mRNA along the way. In bacteria there are two termination mechanisms: the first involves a protein called Rho helicase binding to a cytosine-rich region on the newly formed mRNA. The presence of the Rho protein destabilizes the interactions between the mRNA and DNA, causing them to separate and terminate transcription. The second mechanism involves a guanine/cytosine-rich region of the mRNA, within which base pairs form. The resulting hairpin effectively derails the RNA polymerase, halting transcription.

Translation

By this point I hope it is clear that DNA and RNA are extremely similar molecules, both with sequences spelt out with a similar four-letter alphabet. Proteins, however, are very different beasts, being made up of twenty different amino acids. So translating the information contained in a gene sequence into a protein sequence needs a very different mechanism from the simple base pairing used in DNA to RNA information transfer.

Following the discovery of DNA's structure, establishing the mechanism of translation was one of the most pressing questions of the time. The problem was simple. There are just four bases in nucleic acids. But genes somehow code for the twenty different amino acids found in proteins. So how can a four-letter alphabet be translated into a twenty-letter alphabet? What methods could biology use to get twenty from four?

One of the most notable attempts to resolve this came from Francis Crick. He saw the problem in the following way: The genetic code had to consist of non-overlapping triplets of bases. But if that is the case then how can one triplet be distinguished from the next. After all there is no punctuation in DNA. It is like trying to find three-letter words in SATEATEATS without any spaces. They could be SAT EAT EAT, or ATE ATE ATS, or TEA TEA, depending on where you started.

Crick came up with an elegant theory he called 'codes without commas'. He took the sixty-four possible triplet codes and put them together in groups according to whether they had the same circular permutations (i.e. ACG, CGA, and GAC are in one group; CCG, GCC, and CGC form a second group; and so on). He then hypothesized that only one sequence from each group would be used to code for an amino acid. These he called 'sense' codons. The remainder were termed 'nonsense'. So if ACG and CCG are sense then the sequence ACGCCGACG can only be read ACG CCG ACG, because CGC CGA both give nonsense codons.

Also into the nonsense bin went AAA, UUU, GGG, and CCC because they would cause ambiguity about where a codon started (e.g. is CCCCGGG read CCC CGG or CCC GGG?). Crick's code without commas theory gives us a total of sixty-four possible codons, and removing CCC, GGG, AAA, and UUU leaves us with sixty. Of those remaining, only every third codon is 'sense', neatly leaving twenty codons to code for the twenty amino acids.

The comma-free code was so elegant, and the numbers fitted so well that everyone believed it for the best part of five years. Until, in 1961, Marshall Nirenberg and Johann Heinrich Matthaei produced a stretch of RNA composed of just uracil. When they added it to a mix of ribosomes, tRNAs, and amino acids necessary to initiate translation, the result was a polypeptide of pure phenylalanine. According to Crick's theory poly-uracil should have been a nonsense sequence, and so his code without commas theory was consigned to a dusty shelf of biochemical history.

In fact, the genetic code is much simpler. The sixty-four possible three-letter codons are largely accounted for by the fact that most amino acids (methionine and tryptophan being the exceptions) are represented by more than one codon. For example, GUU, GUC, GUA, and GUG all give rise to valine.

There are also four codons that serve additional functions. AUG codes for methionine, but it is also the start codon, signalling the point in an mRNA sequence where translation is initiated. Meanwhile UGA, UAG, and UAA are stop codons, which indicate where on an mRNA molecule translation needs to be halted. Amazingly this genetic code (Figure 15) is shared (with some rare and minor variations) by all lifeforms on Earth.

It is here that the effects of mutations become apparent. The redundancy in the system can mitigate some changes to a genetic code, for example a protein sequence is tolerant of any mutation that alters the last nucleotide in a valine codon (these are known as silent mutations). Proteins are often robust enough to accommodate changes that result in amino acid substitutions. Nevertheless, in some cases even the single-letter change such as GAG to GTG, which results in a valine in place of a glutamate, can cause dramatic symptoms. When this occurs at a specific point in a haemoglobin gene the result is sickle cell disease.

Translation occurs in large and elaborate RNA-based structures called ribosomes. These archetypal molecular machines are probably another relic from the RNA world. Since then they have evolved considerably, and there is now substantial variation throughout the tree of life. But broadly speaking they consist of

Second letter

		U	C	A	G	
U		UUU ⎫ Phe UUC ⎭ UUA ⎫ Leu UUG ⎭	UCU ⎫ UCC ⎬ Ser UCA ⎪ UCG ⎭	UAU ⎫ Tyr UAC ⎭ UAA Stop UAG Stop	UGU ⎫ Cys UGC ⎭ UGA Stop UGG Trp	U C A G
C		CUU ⎫ CUC ⎬ Leu CUA ⎪ CUG ⎭	CCU ⎫ CCC ⎬ Pro CCA ⎪ CCG ⎭	CAU ⎫ His CAC ⎭ CAA ⎫ Gln CAG ⎭	CGU ⎫ CGC ⎬ Arg CGA ⎪ CGG ⎭	U C A G
A		AUU ⎫ AUC ⎬ Ile AUA ⎭ AUG Met	ACU ⎫ ACC ⎬ Thr ACA ⎪ ACG ⎭	AAU ⎫ Asn AAC ⎭ AAA ⎫ Lys AAG ⎭	AGU ⎫ Ser AGC ⎭ AGA ⎫ Arg AGG ⎭	U C A G
G		GUU ⎫ GUC ⎬ Val GUA ⎪ GUG ⎭	GCU ⎫ GCC ⎬ Ala GCA ⎪ GCG ⎭	GAU ⎫ Asp GAC ⎭ GAA ⎫ Glu GAG ⎭	GGU ⎫ GGC ⎬ Gly GGA ⎪ GGG ⎭	U C A G

First letter (left side) — Third letter (right side)

15. Genetic code.

three or four RNA chains and dozens of scaffolding proteins which together form separate sub-units.

An mRNA strand is fed into the smaller ribosomal sub-unit. Meanwhile, amino acids are transported to the larger ribosomal sub-unit, attached to transfer RNAs (tRNA; Figure 16). There are sixty-one different tRNAs, one for each amino acid coding codon, and each tRNA has a free three-letter sequence (anti-codon) that is complementary to that codon. For example, four tRNAs exist to carry valine, and they have anti-codons CAA, CAG, CAU, and CAC, which are complementary to the codons for valine: GUU, GUC, GUA, and GUG. The ribosome assists the docking of tRNAs to the correct codon on an mRNA, and then holds this in place whilst a second tRNA binds. The two amino acids are now in proximity to each other, allowing the ribosome to catalyse the formation of a peptide bond between them. The first tRNA, now

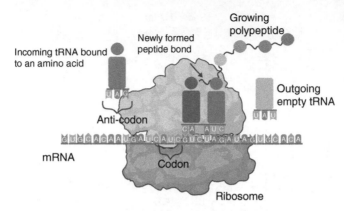

16. tRNAs loaded with amino acids bind to codons on an mRNA within a ribosome. The ribosome catalyses formation of peptide bonds between amino acids to synthesize proteins.

devoid of its amino acid cargo, is released, and the ribosome moves along the mRNA. The process of tRNA docking, catalysis, and shuffling of the ribosome is repeated until it encounters a stop codon, for which there are no tRNAs. Instead proteins called release factors bind here and serve to knock the ribosome off the mRNA, in the process releasing the newly constructed protein into the cytoplasm where it is free to fold and function.

Chapter 5
Powering a cell: bioenergetics

Now that we have encountered all the groups of major macro- and micromolecules that play a role in biochemistry it is time to look at some examples of how they work together to create interconnected pathways of chemical reactions.

The Universe tends towards disorder and life is essentially a battle against this ever-increasing entropy. Combating entropy takes energy. And for almost all life on Earth (directly or indirectly) the source of that energy is the Sun. The biochemical process that collects the Sun's rays and uses their energy to convert carbon dioxide and water into carbohydrate is of course photosynthesis.

Sunlight being such a rich source of energy, not surprisingly, more than one branch of life has evolved light harvesting processes. Red and green algae, purple halobacteria, red photosynthesizing bacteria, and of course green plants all use a variety of means to trap the energy from the Sun's rays. But in all cases photosynthesis starts with a photon of light hitting a pigment molecule. For purple halobacteria that pigment is a variant of vitamin A called retinal (which is also the pigment in light-sensing proteins in our eyes), phycoerythrin give red algae their hue, while orangey-red pigments called carotenoids play a role in plants and green algae. But the most common light harvesting pigments are chlorophylls. There are in fact two different common chlorophyll molecules used by plants (named chlorophyll A and B), but for the sake of brevity I won't make any further distinctions between them.

The light reactions

Life on Earth began some 3.7 billion years ago; it took another 1.2 billion years for the first cyanobacteria to evolve a photosynthetic process that produced oxygen. Chloroplast, the organelles where photosynthesis takes place in plants and algae, are thought to be derived from ancestral cyanobacteria that were engulfed by a primordial eukaryote, so imbuing it with the power of photosynthesis. Traces of those assimilated cyanobacteria are still apparent in a few genes that reside in chloroplast.

Chloroplast are divided into membrane-bound compartments called thylakoids. At the edge of the thylakoid, embedded within the lipid membrane, are two protein complexes called photosystem I (PSI) and photosystem II (PSII). These are the dominant light harvesting complexes on Earth, and the critical role they play in powering our ecosystems becomes strikingly apparently when you realize that the vast green swathes of the Earth visible from space are predominantly due to PSI and PSII.

PSI is so named because it was the first to be discovered; however PSII plays an earlier role in the photosynthetic process, so we will start there. PSII is a huge conglomeration of about twenty protein sub-units, and a hundred or so cofactors including some thirty-five chlorophyll molecules, plus metal ions, lipids, and numerous other pigments.

Photosynthesis starts with a photon of light hitting a single chlorophyll molecule on PSII (Figure 17). The energy of that photon (and hence its wavelength and colour) is critical for what happens next, because the chlorophyll molecules absorb violet and blue at one end of the visible light spectrum along with reds and oranges at the other end. This leaves a great big gap of green and

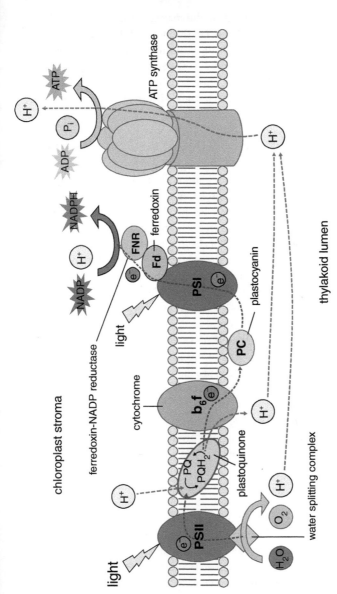

17. The light dependent reaction of photosynthesis taking place in chloroplast thylakoid membranes.

yellow light in the middle of the spectrum which is reflected, hence plants appear green. On the face of it this seems a bit odd, as the Sun's emission actually peaks in the green and yellow range, so plants reject a good chunk of the Sun's energy. Why plants have evolved in this way is not entirely clear. It may be as simple as the fact that black plants, which would absorb all the light from the visible spectrum, might overheat.

Returning to the photosynthetic process, let us assume that a 'blue' photon hits a chlorophyll molecule; the absorption of the photon knocks an electron on the chlorophyll up to a higher energy state. If the chlorophyll were in isolation then the electron would simply return to its ground state, releasing the energy as red light. In PSII the array of pigments are precisely arranged to catch the 'red' energy and act as a chain of electron acceptors which shuttle the excited electrons around the photosystem. From there they are moved through the membrane via a lipophilic molecule called plastoquinone and on to the next protein complex in the light reaction, cytochrome b6f. This however leaves the PSII lacking an electron. Replacement of the electron is facilitated by another region of PSII called the water splitting complex. As the name suggests, it extracts electrons from water, resulting in hydrogen ions (protons) and a waste product, oxygen. It is from this biochemical step that the oxygen in our atmosphere is derived.

Cytochrome b6f transfers electrons to the water-soluble acceptor plastocyanin. In the process a proton is pumped across the membrane and into the thylakoid space.

Meanwhile within the other light harvesting protein complex, PSI, a similar process occurs to that in PSII. Here too light excites electrons on the chlorophyll, but this time the electron is transferred to ferredoxin and through the final link in the chain to an enzyme called ferredoxin NADP reductase. This enzyme uses the electrons to recharge NADPH, which is an essential cofactor used in metabolic processes through plant metabolism, and most notably within the

Calvin cycle, by which carbon dioxide is converted to glucose. After all this is complete PSI is also left with missing electrons, but it is not able to strip them out of water; instead it receives the electron that was deposited on plastocyanin earlier in the process.

Meanwhile, protons from the PSII water splitting complex and the cytochrome b6f have been building up inside the thylakoid space. The result is a higher concentration of protons within the thylakoid space compared to without. This proton gradient is in effect a store of potential energy. And in much the same way water can flow downhill through a turbine to convert potential kinetic energy into electricity, protons flow through another protein called ATP synthase (which even turns like a turbine!), resulting in the conversion of ADP and a phosphate ion (Pi) to ATP.

This whole amazing process, involving four protein complexes and their numerous cofactors, can be summarized in one simple reaction scheme:

$$2H_2O + 2NADP^+ + 3ADP + 3P_i \rightarrow O_2 + 2NADPH + 3ATP$$

As you will have noticed, as yet there has been no sign of carbon dioxide or carbohydrates. That is because the process described so far constitutes just the so-called light reactions, which constitute one-half of photosynthesis. The second part of the process relates to the dark reactions. The dark reactions, also called the Calvin cycle, do not actually take place in the dark. But light does not play a direct role in the reactions, hence the slightly misleading name. It is in this part of the process that ATP and NADPH, created by the light reactions, are used to convert carbon dioxide into the building blocks of carbohydrates.

The dark reactions

All life on Earth is based on carbon chemistry; every protein, lipid, nucleic acid, hormone, and carbohydrate is built on a carbon-based scaffold. The biosphere's main source of carbon is carbon dioxide, and there is just one enzyme that dominates the process of extracting it from the air and feeding it into the ecosystem. That enzyme, called ribulose-1,5-bisphosphate carboxylase/oxygenase, commonly abbreviated to rubisco, sits within the chloroplasts of plants, and it is a linchpin of life. Rubisco stores chemical energy within the high energy electrons of stable bonds between carbon atoms of carbohydrates. Some staggering numbers serve to illustrate just how crucial this enzyme is to life on Earth. Rubisco typically makes up 50 per cent of the protein in a plant leaf, so this means there is probably 4×10^{10} kg of rubisco spread over the globe; together these enzymes fix about 100 gigatonnes of carbon per year—a mass fifty times that of the entire human population.

But despite its critical role in our ecosystem, rubisco is rather inefficient as an enzyme. It can handle just a few carbon dioxide molecules per second (compare this to catalase's turnover rate (k_{cat}), which is in the order of 1,000,000 per second). Not only that but it often makes mistakes; it is prone to using oxygen instead of carbon dioxide as a substrate. When it does so this leads to the formation of a toxic compound called 2-phosphoglycolate, which the plant then needs to put in a significant effort to dispose of. The reason why rubisco is quite so inefficient and non-specific may be that it has fallen into an evolutionary trap. When photosynthesis first emerged as a biological process, atmospheric oxygen levels would have been vanishingly low, so there was no evolutionary pressure for plants to discriminate between carbon dioxide and molecular oxygen. In short, the toxic side reaction would not have been an issue. However, as atmospheric oxygen levels increased, the toxic reaction began to have an effect; plants churned out more and more 2-phosphoglycolate, and effectively poisoned themselves. In response rubisco had to evolve to be more specific for carbon dioxide. There is generally an inverse correlation between enzyme specificity and speed, so as rubisco evolved to discriminate between the two gases its catalytic activity, in turn, slowed down. Now, hundreds of millions of years later, modern plants are still dealing with their own 'toxic' waste, but evolution has provided a trade-off, with lesser efficiency in exchange for greater specificity.

Rubisco, like PS1 and PSII, is another huge protein complex, and in plants it is constructed from eight identical copies of a large protein sub-unit (which is encoded by the chloroplasts' own DNA) and eight identical copies of a smaller sub-unit (encoded by the nuclear DNA); together they form a ball-like structure, with an atomic mass of 540,000, some 12,000 times bigger than the carbon dioxide molecules that it extracts from the air. Rubisco's CO_2 binding site is formed from eight amino acid side chains and a magnesium ion, which serve to hold the CO_2 in place long enough for it to be added onto a five carbon compound called ribulose 1,5-bisphosphate (RuBP). The product of this reaction is an unstable six carbon compound; 3-keto-2-carboxyarabinitol-1, 5-bisphosphate, which almost instantly falls apart to leave two three carbon molecules of glycerate-3-phosphate (G3P).

The remainder of the dark reactions are simply there to regenerate the RuBP substrate for rubisco. This takes place in a number of steps, which sees most of rubisco's own G3P product along with ATP and NADPH formed during the light reaction being fed into the Calvin cycle to produce fresh ribulose 1,5-bisphosphate for rubisco to once again react with another carbon dioxide molecule. The stoichiometry of the cycle means that for every six G3Ps (eighteen carbons) produced by rubisco, five of them are put back into the cycle to produce three RuBPs (fifteen carbons). This means that the cycle has to be completed three times, at a cost of nine ATPs and six NADPH molecules, to manufacture just one spare G3P. From here on the metabolic processes rapidly diverge. Some of this small excess of G3P is kept in the chloroplasts, where two molecules are combined to form glucose. This glucose may, amongst other things, end up as nucleotide derivatives or be stored as starch. Meanwhile, other

G3Ps are exported from the chloroplast and transformed into fructose-6-phosphate, sucrose, and much more besides.

The upshot of photosynthesis is that plants convert light and carbon dioxide into sugars and carbohydrates. Other organisms, including ourselves, then utilize the stored chemical energy in those carbohydrates, and in the process release carbon dioxide back into the atmosphere.

Glycolysis

At first the oxygen produced by photosynthesizing cyanobacteria had little impact on the atmosphere; most of the oxygen simply caused iron in the Earth's crust to rust. However, after 200 million years this oxygen sink had become saturated and as a result biochemistry made its first major impact on the very geology of our planet. With nowhere else to go the oxygen accumulated in the oceans and atmosphere and so life altered the chemistry of the planet's air and seas as well. Life quickly adapted to these new conditions, and in just another 600 million years organisms that thrived on the waste oxygen produced by photosynthesis were widespread. The ability to utilize oxygen kickstarted the explosion of multi-cellular life on Earth. Prior to this, life, in relatively simple forms, had managed without oxygen for almost two billion years.

The remains of the biochemical process that limited life on the pre-oxygenated Earth still forms the start of the metabolic pathway that converts glucose into the chemical energy currency of the cell—ATP. That process is glycolysis, a ten-step pathway, each stage of which is catalysed by a specific protein. The outcome of glycolysis is a paltry two molecules of ATP, two of NADH, and one of pyruvate. The ATP represents just a fraction of the potential energy locked up in glucose (most of which is still tied up in the pyruvate), but in the absence of oxygen there is not much more a cell can do, beyond regenerating a molecule of NAD+

which feeds back into the glycolysis pathway. Animals and some bacteria (such as *Lactococcus lactis*, the micro-organisms mentioned at the start of this book) achieve this by converting the pyruvate to lactate via the enzyme lactate dehydrogenase. Meanwhile yeast bolts two steps on to the end of the glycolysis pathway, creating the ethanol fermentation process. This consists of conversion of pyruvate to acetaldehyde and carbon dioxide (hence the bubbles in beer), followed by the action of an enzyme called alcohol dehydrogenase, which catalyses the formation of ethanol from the acetaldehyde.

Ethanol dehydrogenase also crops up in organisms which do not ferment ethanol, including ourselves. In these cases its purpose happens to illustrate an interesting feature of enzymes: they can work both ways. In yeast the enzyme converts acetaldehyde to ethanol, but when we drink alcohol, a similar version of the enzyme converts ethanol to acetaldehyde. (It is this acetaldehyde, which is much more toxic than the ethanol, that contributes to a hangover.)

Accessing the rest of the energy from glucose requires an oxidizing agent. Before the accumulation of oxygen in the atmosphere there was no freely available chemical to do this job. Just as fire was absent without free oxygen in the atmosphere, so too organisms were unable to fully 'burn' glucose. So life was limited to simple unicellular organisms. But all that changed following the so-called Great Oxygenation Event, and as oxygen levels slowly built up, new biochemistry evolved to take advantage of the chemical opportunities afforded by oxygen's reactivity.

The last inefficient anaerobic steps were replaced by an elaborate process all of which takes place in the powerhouse of cells, the mitochondria. These membrane-bound organelles, found in all eukaryotic life, which, like chloroplasts, originate from free-living bacteria, specialize in the biochemistry of extracting potential energy from carbohydrates (and fats, but for the sake of brevity we won't explore those pathways here). The mitochondrial membranes are much more than just a means of containing the reactions; as we will see, they play critical roles in the whole process. In fact, huge areas of membrane are needed, which gives rise to highly creased and folded inner mitochondrial membranes. These in turn are enfolded in a smooth outer mitochondrial membrane. The outer membrane is fully permeable to metabolites, so it does little more than hold the inner membrane in place. However, the inner membrane is much more selective. This creates the first challenge: glycolysis happens in the cytosol (the volume within a cell but outside of organelles) of cells, so before any more manipulation can take place pyruvate needs to be transported into the mitochondria. This was clear from the moment that the role of the mitochondria became apparent. Nevertheless it took forty years of searching before (in 2012) the mitochondrial pyruvate carriers were finally identified.

Once pyruvate has gained entry into the mitochondria, extracting the remaining chemical energy from glucose can start in earnest. First pyruvate is converted to acetyl-CoA. This is fed into a circular reaction scheme, known as the citric acid cycle (or Krebs cycle, after Hans Adolf Kreb who, along with William Arthur Johnson, identified the cycle in 1937). Within this cycle acetyl-CoA is bound to oxaloacetate to form citrate. Another seven steps (making eight in total) complete the cycle and regenerate the oxaloacetate ready for another turn of the wheel. In the process pyruvate is disassembled into carbon dioxide and water. The biochemical point of all this are the four NADH, reduced flavin adenine dinucleotide ($FADH_2$), four protons, and a molecule of GTP (which has a similar function to ATP) generated by one turn of the cycle. All of these are trapped within the mitochondrial matrix, and, for what happens next, this location is as important as the molecules themselves.

The final stage of the process is linked to Peter Mitchell's chemiosmotic theory. In 1961 Mitchell described the fundamental energy conversion mechanism in biochemistry (although at the time it was highly controversial as it ran contrary to the theories of the day). It essentially states that captured energy is used to move protons across a membrane (we have touched on this in exploring the dark reactions of photosynthesis). This generates a high concentration of protons on one side of the membrane and a low concentration on the other. The movement of protons back across the membrane through the ATPase complex drives the formation of ATP. And this is why the membrane is so important, because it forms the barrier between the high and low concentrations of protons.

One more component links the citrate cycle and the ATPase: the proteins in the electron transport complex. These proteins accept the electrons being carried by the NADH and $FADH_2$. The electrons are in a high energy state, and the proteins control their fall to lower energies. As they do so they channel that energy to pump protons across the inner membrane and out of the mitochondrial matrix. This generates the proton motive force used to drive ATPase and the production of ATP. All together the aerobic mitochondria churn out a net thirty-two ATPs (once the cost of transport is taken into account) from a glucose molecule, a vast improvement on the two from glycolysis.

I started this section describing the importance of oxygen. However it won't have escaped your notice that, so far, oxygen has not featured in any of the reactions. Indeed its role is so simple that it is easy to overlook it, amongst the complexity of the cycles, transporters, and so on. Electrons on the final electron transport complex need somewhere to go, otherwise a backlog forms in the chain and the whole mitochondrial powerhouse grinds to a halt. Oxygen acts as that final electron acceptor, its electronegativity making it ideally suited to the task, and since it is extrinsic to the organism, and plentiful in the atmosphere, there is a ready supply of it. So the final reaction in the complexity that is aerobic respiration is quite simple: an oxygen molecule (O_2), plus four electrons and two protons combine to form two molecules of water.

Chapter 6
Manufacturing and maintaining DNA

DNA replication

Upon deducing the structure of DNA, Watson and Crick stated, 'it has not escaped our notice that the specific pairing we have postulated immediately suggests a possible copying mechanism for the genetic material'. They were alluding to their observation that the double helix can be separated into individual strands, with each serving as a template on which new strands can be assembled. This simple statement doesn't even begin to capture the astronomical task faced by organisms as they rapidly, accurately, and completely replicate their DNA. But a back of the envelope calculation might:

The distance between base pairs in DNA is 0.34 nanometres. The human genome is comprised of about three billion base pairs, so each cell contains about 1 metre of double-stranded DNA. It is estimated that a human body is made up of thirty-seven trillion cells, which means there is a total of 37 billion kilometres of DNA within your body—enough to stretch from the Sun and to Neptune and back four times. And if that is not remarkable enough, a cell reproduces an average of 300 times, giving a total of over 11 trillion kilometres of DNA manufactured in a lifetime. If you were to extend that out in one long line it would take fourteen months for light to traverse from one end to the other. In other words, before you die, your body will probably manufacture more than a light year of DNA!

What is all the more incredible is that your biochemistry manages to make all that DNA with astonishingly high fidelity. It generally makes just one mistake every billion base pairs. That is the equivalent of typing out copies of the complete works of Shakespeare and making just one typographical error in every 2,000 copies.

Before we explore the marvellous examples of molecular machinery that make all this possible, it is worth considering the mechanism that Watson and Crick were referring to. They had spotted that the double helix of DNA could easily be separated into two individual strands, each of which acts as a template for the manufacturing of a new complementary strand. This became known as semi-conservative replication, to indicate that each daughter DNA double helix contained one new strand and one old strand, conserved from the parent molecule.

The proof of this had to wait five years after the publication of Watson and Crick's double helix structure, for an elegant experiment designed by Matthew Meselson and Franklin Stahl (Figure 18). They grew bacteria in a medium containing a heavy isotope of nitrogen (^{15}N). The bacteria incorporated this into their DNA. Meselson and Stahl then switched the growth medium to lighter nitrogen (^{14}N) and allowed the bacteria to grow for just one generation. When they extracted the DNA from these bacteria and checked its density, they found that it was exactly intermediate between pure ^{15}N DNA and pure ^{14}N DNA. This indicated that each double strand contained a 50:50 split of old and new DNA. However, this did not prove a semi-conservative replication process over dispersive replication, by which each daughter strand would contain a mixture or old and new DNA, as both would have produced this result. So Meselson and Stahl

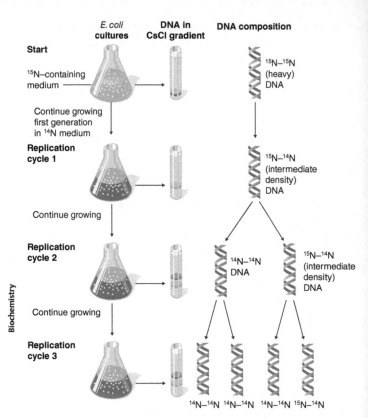

18. The Meselson and Stahl experiment, which demonstrated semi-conservative replication of DNA.

allowed the bacteria to grow for a second generation in ^{14}N medium. This resulted in two populations of double-stranded DNA, one containing two strands of ^{14}N DNA (one newly replicated and the other from the parent DNA) and another containing a mixture of ^{15}N DNA (from the parent DNA) and a newly replicated ^{14}N DNA strand. The only way that this could come about was if each DNA strand stayed intact during the replication process and was passed on to the next generation, thus proving semi-conservative replication.

The concept underpinning semi-conservative replication is quite simple; all it requires is for the two strands of the DNA helix to be separated and the resulting single strands copied. But the mechanism by which it occurs is much more complicated. The feat is coordinated by a group of about twenty proteins, working in a concerted fashion and collectively known as the replisome. The process in eukaryotes and prokaryotes differs slightly, but to illustrate the mechanism we will just take a look at the simpler *E. coli* prokaryotic system.

Bacterial genomes are circular, so the first step is to establish where on this ring to start. This occurs at a particular DNA sequence called the replication origin. About twenty copies of a protein called DnaA bind to this region and together they prise apart the double strands, creating a loop of single-stranded DNA sandwiched by two double helical sections. DnaA's powers are limited: it can only hold this region open, and is unable to propagate the opening. Unwinding the rest of the DNA is the job of a helicase enzyme called DnaB. It binds to the single-stranded DNA created by DnaA and then, powered by ATP, it works its way along the single strand, extruding it through the centre of its barrel-like structure. As a result, the complementary strand is forced to the outside of the barrel, and so DnaB effectively acts as a wedge driving itself between the two strands. Meanwhile, in its wake, single-stranded DNA binding proteins latch on to the newly separated DNA ensuring it does not anneal straight back together again. One more thing needs to be in place before the actual copying process can get underway. Another enzyme, RNA primase, creates small sections of complementary RNA, and it is from these that the DNA polymerase starts work creating nascent DNA strands.

There are in fact at least five DNA polymerases in *E. coli*, three of which are involved in the normal replication process. DNA polymerase III (Pol III), a huge complex constructed from seventeen sub-units, has the ability to power through almost

1000 nucleotides per second and so does the lion's share of the work. DNA polymerase I (Pol I) and II (Pol II) are much smaller, single sub-unit enzymes. Pol II is involved in repairing damage and acts as a backup polymerase. Meanwhile Pol I replaces RNA primers with DNA as well as having a role in proofreading. DNA polymerases IV (Pol IV) and V (Pol V) only come into action if there is a problem with the replication polymerases. Damage in DNA can cause lesions, like buckled railway lines, that cause Pol III to stall at the replication fork. Pol IV and Pol V play a role repairing the fault and getting the Pol III train moving again.

All of these polymerases catalyse fundamentally the same reaction, the addition of a deoxynucleotide (dNTP) to the 3' end of nucleic acids (DNA or RNA). Several components need to come together within the polymerase active site to make this happen: the template DNA, a 3' end of a nucleic acid, a free dNTP, and two magnesium ions.

The positive magnesium ions form electrostatic interactions with the phosphates on the dNTPs and the aspartate side chains within the enzyme active site (Figure 19). This helps to stabilize the base pairing of the dNTPs with the template DNA. One of the magnesium ions also helps to deprotonate the 3'-OH which can then form a covalent bond with the first phosphate of the nucleotide. In the process the remaining two phosphates are liberated, and the polymerase moves onto the next base on the template DNA.

In many ways DNA polymerases are rather unusual enzymes, because their active site can accommodate four separate reactants, that is, any of the correct base pairs; C-G, C-G, A-T, and T-A. Not only that but incorrect pairings are geometrically excluded from the active site, massively reducing the chances of a mismatch in the product DNA.

At this point things get complicated, because DNA polymerase has a serious limitation. It can only move in one direction (from

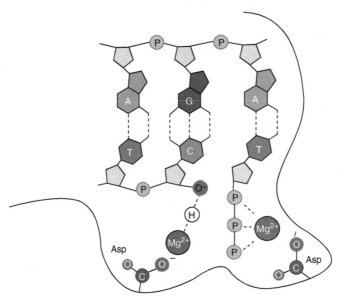

19. DNA polymerase active site.

the 5' to the 3' end) along a strand. For one (leading) strand this is no problem, the polymerase just starts off at its RNA primer, and then chases the ever-moving replication fork being generated by the helicase, churning out a single continuous complementary strand of DNA as it goes. Meanwhile, on the other (lagging) strand, the RNA primer is generated intermittently by an RNA primase attached to the back of the helicase. DNA polymerase starts at these and moves away from the replication fork until it comes across the back end of a previously made fragment. The polymerase then falls off and rebinds at the replication fork (Figure 20). This apparently cumbersome mechanism results in a series of disconnected sections (called Okazaki fragments) which need stitching together by yet another enzyme called DNA ligase.

Along with the polymerases, helicases, ligases, and so on, there are yet more proteins within the replisome doing such jobs as

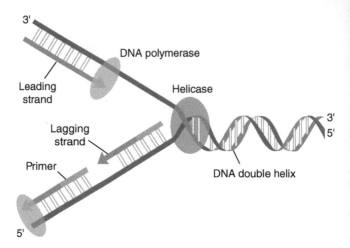

20. The DNA replication fork. Helicase unwinds the double helix, whilst DNA polymerases create new strands of DNA. On the leading strand it does this continuously, on the lagging strand it does so in a discontinuous fashion.

clamping polymerases to the DNA, and even structures that load the clamps onto the polymerases. I won't describe these in detail, but I would like to introduce you to one more protein in the replisome, if for no other reason than that I find its ability to manipulate multiple strands of DNA rather wonderful. It is called DNA gyrase and its job is best illustrated with a demonstration.

If a long elastic band or loop of string is held taut between two fingers and rotated to produce five or six twists in the band, the coiled section roughly resembles a stretch of double-stranded DNA. If now a thumb is inserted within the loop and moved towards one end, pushing the coiled section as it moves, the thumb is doing the same job as a helicase, creating single-stranded sections behind it. But it is evident that it is not removing the twists, just forcing them together, creating supercoils and tension further up the band. Helicase causes the same thing to happen in

DNA, and that tension is relieved by DNA gyrase. To achieve this the DNA gyrase cuts right through the double helix, attaches to a second section of DNA, and passes that through the gap. Finally, it repairs the cut, leaving intact DNA with a twist removed. And it repeats this over and over again, all the time keeping ahead of the DNA helicase.

High fidelity

I mentioned earlier that DNA replication occurs with astonishingly high fidelity, making one mistake every ten billion bases or so. However, DNA polymerases alone are not so accurate. Once every 100,000 nucleotides or so they make a mistake, and so built into every Pol III complex there is a proofreading function.

An incorrectly inserted nucleotide causes a distortion in the DNA, and the protein can detect this slight conformational irregularity. When it does so, the polymerase stops and goes into reverse. It then swings the end of the nascent chain into a second active site. Here an exonuclease excises the offending nucleotide. The strands are then placed back in the polymerase region and the whole enzyme complex moves forward again. This simple bit of editing reduces the error rate considerably, to about one in ten million base pairs, but that still is not quite good enough to maintain the integrity of a genome.

An additional check also takes place shortly after replication. Errors at this stage are, however, much more difficult to detect. When a mismatch happens during replication the DNA polymerase enzyme has the template in hand, and so it 'knows' where the error lies, that is, at the end of the newly synthesized strand. But if a mismatch gets past the replisome, how is a cell going to tell which strand contains the correct sequence and which holds the error? The simple answer is that *E. coli* labels its DNA strands. However, this labelling takes a few minutes. So, in that time window cells are able to distinguish parent from daughter strands. This allows a mismatch repair system to check the daughter strand for errors.

The labelling task falls to DNA adenine methylase, which attaches methyl groups (-CH_3) onto adenines—but only when they occur in the sequence GATC. That sequence is important, because its complementary sequence is CTAG—the same but in reverse. This means that once DNA adenine methylase has completed its run through the newly synthesized genome (which takes about two minutes), there will be numerous pairs of methyl groups virtually opposite each other. The GATC sequence crops up fairly regularly in the *E. coli* genome, so the methyl groups adorning them act as regular signs flagging up the older, parent strand. Or at least they do for those two minutes, and that window of opportunity is plenty long enough for the mismatch repair system to do a sweep through the genome.

The mismatch repair system consists of three proteins. The first to be activated are MutH (Figure 21). These bind to the GATC/CTAG sites but only when there is a single methyl group present, that is, one of the DNA strands has been newly synthesized. Next, MutS scans between pairs of MutH and then binds to mismatch mutations. Once MutS is locked into place it recruits the third protein, MutL. DNA is then dragged, from both directions, through the MutL-MutS complex, creating a loop of DNA out of the top. In the process the two MutH proteins eventually collide with the MutL-MutS and together they form an even bigger complex. This then triggers an endonuclease function in the MutH, each of which puts a nick in the new DNA strand. The proteins dissociate, bringing the excised strand of daughter DNA with them. Shortly after, DNA Pol III creates a new daughter strand which is stitched into place by a DNA ligase, to complete the repair.

This represents the last chance *E. coli* has to correct any of its own DNA replication errors. However, there is still much that can go

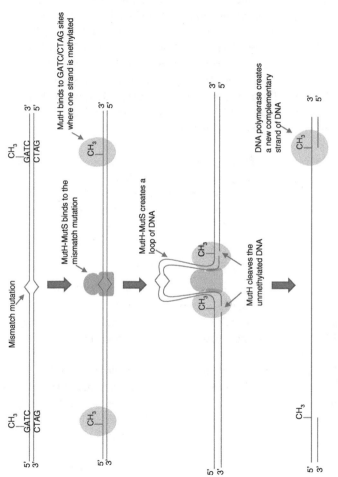

21. MutH mismatch repair system.

wrong with the DNA. For instance, a particularly familiar source of damage is ultraviolet (UV) light. One of the ways it alters DNA is by catalysing the formation of covalent links between adjacent thymines or cytosines. These dimers disrupt the normal base pairing and cause a lesion in the otherwise tidy helical structure, which if left uncorrected has serious consequences for the cell. Without the energy from the UV light, formation of these dimers is highly unlikely. Once formed, they are difficult to undo without a similar energy source. Rather poetically, one of the ways by which the cellular machinery deals with these dimers is with the photolyase system, which utilizes the energy in visible light to undo the damage caused by the UV.

This is just one of the myriad mechanisms the cell uses to carefully maintain the integrity of its precious DNA records. There are many more systems which have evolved to manipulate DNA in order to fix breaks, the result of chemical damage, and even to directly attack the DNA of invading organisms.

DNA sequencing and amplification

Discovering the ways in which cells replicate their DNA led to the question: Can we harness these amazing natural machines to shed light on DNA itself? Two applications, both of which utilize DNA polymerase, have arisen from just such an approach.

The first is DNA sequencing, which has enabled us to read genomes and much more besides. There are now multiple methods for sequencing DNA but the first technique to be widely used was developed in 1977 by Fredrick Sanger (who also first found a way to sequence proteins: see Chapter 1).

Sanger's sequencing begins with a stretch of DNA of interest (Figure 22). First it is denatured by heating it to about 90°C. This causes the double helix to unravel and separate into two individual strands. Then a short single-stranded primer, designed specifically

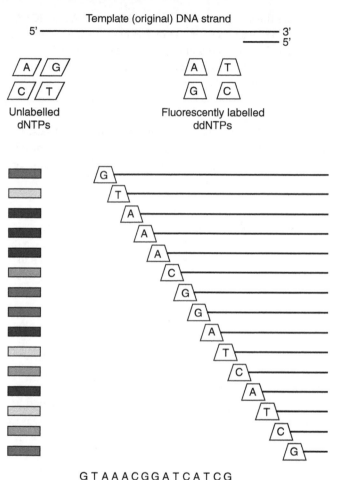

Template (original) DNA strand

Unlabelled dNTPs

Fluorescently labelled ddNTPs

GTAAACGGATCATCG

22. Sanger's DNA sequencing.

to bind to the complementary section of the DNA of interest, is added to the mix. The primer, together with the longer strand, creates a short double-stranded section where a DNA polymerase can start its replication activity. Along with the DNA and the enzyme are the four nucleotides (dATP, dTTP, dGTP, and dCTP, collectively known as dNTPs). At this point the sample is divided into four, and a small amount of a nucleotide variant, called ddNTP, is added to each of the four. This variant differs from dNTPs in one important respect: it is missing the 3'-OH which is needed for the addition of a subsequent nucleotide. Consequently, they can be incorporated into the DNA, but once in place elongation of the DNA strand is terminated. The ddNTPs used in sequencing have one more important characteristic: they include a label of some sort (originally a radioactive label, but now more frequently a fluorescent label).

To illustrate how this allows us to sequence DNA let us imagine what would happen with the sequence GAT TAC AGA TTA C in the tube containing ddATP. The DNA polymerase starts down the sequence making complementary DNA. First it adds a dCTP to complement the G, then a dTTP to match the A. When it arrives at the first T it tries to add a complementary dATP—if it does so it just carries on. But if it attaches a ddATP instead, then the elongation reaction stops. Each time the polymerase reaches for a dATP there is a chance it grabs a ddATP, which causes the reaction to terminate. This means that the tube will contain a mixture of the products CTA, CTAA, CTA ATG TCT A, CTA ATG TCT AA, and CTA ATG CTT AAG. These different length products are then separated out by size, and visualized with the radioactive or fluorescent label. Analysing these data allows us to work out that the third, fourth, tenth, and eleventh bases are all adenines. Meanwhile, the same thing is repeated, in three other separate reactions, with the other dNTPs, and so together the whole DNA sequence is revealed.

Sanger's DNA sequencing earned him a second Nobel prize in 1980 (making him one of an elite group of only four people to have received two Nobel prizes, the others being Marie Curie, Linus Pauling, and John Bardeen). The technique was undoubtedly a revolutionary breakthrough, for it finally allowed us to probe the composition of genes. However, it had a significant limitation. Most obviously it needs a substantial sample of DNA to sequence. At the time, determining the genetic code of a trace sample collected from a few cells, or harvested from fossilized remains, was unimaginable. Realizing these sorts of dreams had to wait for, arguably, the single most important biotechnological development of the modern era: the polymerase chain reaction (PCR; Figure 23).

PCR is so fundamental to modern biochemistry and molecular biology that the *New York Times* described it as 'virtually dividing biology into the two epochs of before PCR and after PCR'. And indeed anyone who has worked with DNA in a lab will have come across this technique (I certainly spent much of my PhD studies using it). PCR's great advantage is that it allows us to create unlimited copies of a sequence of DNA, starting from the tiniest of samples. And this has enabled us to study everything from ancient DNA fragments extracted from Neanderthal bones and frozen mammoths, to a sample taken from an individual for COVID-19 infection, to evidence left at the scene of a crime.

PCR is actually remarkably simple, so simple in fact that its inventor Kary Mullis (who sadly died whilst I was writing this chapter) marvelled, during his 1993 Nobel prize acceptance speech, that no one had thought of it before:

> 'Dear Thor!,' I exclaimed. I had solved the most annoying problems in DNA chemistry in a single lightning bolt. Abundance and distinction. With two oligonucleotides, DNA polymerase, and the four nucleoside triphosphates I could make as much of a DNA sequence as I wanted and I could make it on a fragment of a specific size that I could distinguish easily. Somehow, I thought, it had to be an illusion. Otherwise it would change DNA chemistry forever. Otherwise it would make me famous. It was too easy. Someone else would have done it and I would surely have heard of it. We would be doing it all the time. What was I failing to see?

Original DNA strand

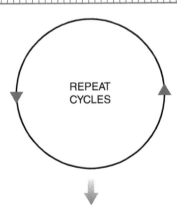

1st cycle   Denaturation

1st cycle   Primer annealing and elongation

Primer   Primer

1st cycle   Cycle completed: 2 copies

REPEAT CYCLES

Multiple copies of DNA

23. The polymerase chain reaction.

Biochemistry

And indeed a generation of biochemists were probably left wondering why they hadn't thought of it themselves. Some even argued (in court no less) that they had; seventeen years before Mullis's work, a group, led by Nobel Laureate H. Gobind Khorana, published something that looks remarkably like PCR.

The story goes that Mullis, whilst driving home from work, was hit by a brainwave (the quote above is part of his recounting of that story). He imagined a way of taking a single copy of a gene and amplifying it a millionfold. First, he would heat the DNA to over 90°C, causing the double helix to denature, just as in the Sanger sequencing technique. Then two DNA primers could be added to the mix, each designed to bind to a different strand of DNA, together bookending the gene. When the mixture was cooled the primers would anneal to their target strands of DNA. If DNA polymerase and plenty of spare nucleotides were in the mix, then the enzyme would bind to the primers and start replicating DNA. At the end of the process, he realized, the number of DNA molecules would have doubled. And if he repeated the cycle then the doubling would occur again and again and again. Do this twenty times and he could create over a million copies of the gene from a single DNA template.

And it worked. Although there was a fly in the ointment: the 90°C needed to denature the DNA also inactivated the DNA polymerase. Consequently, after each heating step of every new cycle, fresh, expensive enzyme had to be added into the mix. The breakthrough that made PCR the ubiquitous technique that it is today came from a bacterium, named *Thermus aquaticus*, which thrived in a scaldingly hot geyser within Yellowstone National Park. Being a thermophile, *T. aquaticus* has a whole complement of thermostable proteins, include a DNA polymerase evolved to work at well over 70°C. When isolated and used in a PCR reaction, this *Taq* polymerase (as it became known) survived the near-boiling temperatures and so negated the need to replenish the enzyme between cycles. As a result, a reaction tube could be set up in minutes, sealed, and then placed in an automatic thermocycler. An hour or so later, the scientists would return to find millions of freshly minted copies of a gene or any other sequence of interest.

So PCR and DNA sequencing make perfect bedfellows. The major limitation of chain termination sequencing, its need for large amounts of DNA, was overcome by the simplicity of PCR. And together they kick-started the boom in biotechnology.

Chapter 7
Following biochemistry within the cell

Biochemistry is fundamentally the study of biological molecules and their interactions. The approach that has dominated the field, ever since Anselme Payen discovered amylase in 1833, is to extract, purify, and isolate sufficient amounts of a molecule so that it became detectable and measurable by the analytical techniques of the time. As a result, for most of the history of biochemistry, we have been studying the behaviour of crowds of molecules. Mulder (Chapter 1) worked out the average elemental composition of egg white; Anfinsen (Chapter 3) watched the result of an ensemble of protein molecules all refolding in their individual ways; and Franklin (Chapters 1 and 4) determined the structure of, not a single DNA double helix, but the mass of molecules packed in her crystals.

This approach, whilst massively productive, has some clear drawbacks. Most notably, by studying the ensemble, we only see average behaviours, and by removing molecules from their native environment we overlook critical interactions.

Instead a preferable approach would be to study individual molecules. When we apply this individualistic methodology to biochemicals we enter the realms of single-molecule biophysics. The great advantage is that, through multiple measurements, we can still derive average behaviours. And we can also elicit information on the distribution of behaviours that sit either side of the mean. As idealistic as this sounds there are of course formidable technical challenges associated with single-molecule studies, not least of which is the signal-to-noise problem.

Patch clamping

Whilst most biochemical techniques were focusing on measuring ensembles of molecules, Erwin Neher and Bert Sakmann (in the 1970s and 1980s) were developing a way to study the function of single-protein molecules. They approached the signal-to-noise problem by isolating an individual protein whilst simultaneously keeping it in its native environment. They achieved this by recording the movement of ions through single-protein channels embedded on the surface of cells. At the time it was known that ions could quickly move across cell membranes; however there was no clear understanding of the mechanism by which cells controlled the flow. Neher and Sakmann reasoned that the movement of ions is essentially the flow of electrical current, therefore measuring the current would provide valuable information on how and what was allowing the ions to flow.

The set-up consisted of a tiny glass micro-pipette with a tip just a micrometre in diameter. The shaft of the micro-pipette contained an electrode and electrolyte. The tip of the micro-pipette was carefully brought into contact with a cell; and a tiny amount of suction was applied to form a seal and firmly clamp onto the cell. The patch within the micro-pipette, the micro-pipette itself, and a second electrode were immersed in a bath of electrolyte, and the two electrodes connected. Together, the electrodes, the electrolyte, and the patch of membrane formed a circuit around which the flow of electrical current could be measured. If the patch of membrane clamped within the micro-pipette contained a solitary channel then the only route for ions to take was through that single-protein molecule.

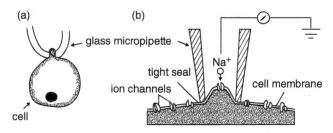

(a)   (b)

glass micropipette

tight seal   Na⁺

ion channels   cell membrane

cell

24. Patch clamping.

When the channel within the isolated patch opened and closed,
Neher and Sakmanns' apparatus clearly showed discrete, tiny
(picoamp) changes in the current (Figure 24). These measurements
conclusively demonstrated that specific proteins control the flow
of specific ions. For example, there are channels that allow sodium
ions to pass, but the passage of very similar ions such as potassium
or calcium is totally blocked by that particular protein, and
instead they must pass through their own bespoke channels.

Patch clamping provided insights into the protein machinery that
controls the flow of ions in and out of cells. It demonstrated that
membrane proteins don't just act as ion-specific holes in a
membrane; some function as gates, only opening in response to
very specific stimuli. The first target studied in any detail was the
acetylcholine receptor, a protein expressed in muscles that detects
the signal from a motor neuron. Acetylcholine is a
neurotransmitter (a chemical messenger released from nerve cells
as a signal to other cells, such as neurons and muscle cells); motor
neurons release acetylcholine at the junction between the nerve
and the muscle cell.

Patch clamping measurements show that without acetylcholine
the receptor stayed firmly shut. However, when the
neurotransmitter was added to the mix the behaviour of the
receptor changed markedly. Surprisingly, it didn't just swing from

a closed to open state, instead it oscillated between the two states, opening for a few milliseconds before shutting again, sometimes for hundreds of milliseconds. This is a perfect example of how a behaviour that is immediately apparent from observations of a single molecule would have been entirely overlooked by an ensemble approach. Studying the receptors *en masse* would have simply shown that as a group, the receptors opened in the presence of the acetylcholine, completely missing the stochastic and binary opening/closing response of the protein.

Green fluorescent protein

In 2008, Osamu Shimomura shared the Nobel prize in Chemistry for work on a protein which revolutionized how biologists and biochemists tracked processes inside single cells.

Shimomura's path to the prize was far from typical. In 1945, at the age of 16, whilst working in a factory repairing aircraft, he witnessed the blinding flash of light and the pressure wave emanating from the nuclear blast over Nagasaki, its epicentre just 12 kilometres away. In war-ravaged Japan there were few educational opportunities; nevertheless by 1951 Shimomura had graduated with a degree in pharmacy (despite having no intention of being a pharmacist), and obtained a placement in a chemistry lab at Nagoya University. His supervisor was interested in bioluminescence, and in particular what made a species of mollusc glow. This was a seemingly impossible task for a young, inexperienced student; other groups of established scientists had failed to make headway despite decades of work. Nevertheless, after just ten months of effort, Shimomura succeeded in isolating and crystallizing a luciferin molecule. This formed the substrate for a luciferase enzyme which catalysed a reaction that emitted light, causing the mollusc to glow. His supervisor was so impressed he arranged for Shimomura to be awarded a PhD despite not being enrolled as a doctoral student. The work also made an impression much further afield, catching the eye of

Professor Frank Johnson at the University of Princeton, who offered Shimomura a post researching another naturally green luminous organism, the jellyfish *Aequorea victoria*.

During the summer of 1961, Johnson and Shimomura collected thousands of jellyfish and extracted a protein that absorbed blue light and emitted it as green, which they named Green Fluorescent Protein, or GFP. GFP proved to have a particularly useful and hitherto unseen feature. All previously studied bioluminescent systems, and other proteins that respond to light, required additional ligands or cofactors to act as a chromophore that absorbs and releases the light. For example, the system on which Shimomura 'cut his teeth' emits light when a luciferin molecule is oxidized by the luciferase enzyme, while the light harvesting proteins discussed in Chapter 5 need chlorophylls to act as light antennas. Not so GFP. It is a self-contained luminescent system, formed from amino acids alone, without the need for an extrinsic chromophore. GFP's intrinsic chromophore is formed from three neighbouring amino acids, a serine or threonine, tyrosine, and glycine (at positions 65, 66, and 67). These are forced into a particular arrangement by the protein's fold. Then, in the presence of oxygen, the threonine and glycine react to form an unusual cyclic structure. And this, along with the neighbouring tyrosine, forms the chromophore (Figure 25(a)).

In 1988, Martin Chalfie heard about GFP and realized that its unique features could help him study the location of other proteins in his (and many developmental biologists') favourite model organism: the tiny, transparent worm *Caenorhabditis elegans*. Chalfie was interested in identifying where genes for a touch receptor were expressed in *C. elegans*. So he fused the GFP gene to a gene promotor that normally activates those receptors. The result was published in the journal *Science* in February 1994. On the cover of the issue is a startling image showing a faintly green worm, and within it several bright green neurons are clearly visible (Figure 25(b)), demonstrating, as the final line of the

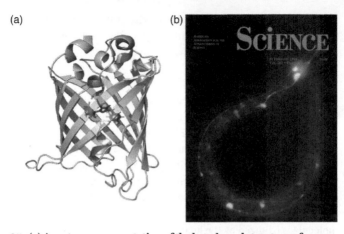

25. (a) A cartoon representation of the beta-barrel structure of green fluorescent protein. In this diagram two sections of beta-strands have been made transparent to better reveal the chromophore (highlighted in dark grey and in a stick representation) within; (b) the cover of *Science* from 11th February 1994, featuring a *C. elegans* worm expressing GFP in its touch receptors.

paper's abstract succinctly put it, 'GFP could be used to monitor gene expression and localisation in living organisms.' For this, and further developments with GFP, Chalfie shared Shimomura's Nobel prize.

This was just the beginning of the applications for GFP. It quickly became clear that GFP could also be genetically fused directly to other proteins, without impacting the function of either protein. This meant that GFP could be used to track proteins within organisms and individual cells.

The final third of the Nobel prize went to Roger Tsien, who studied the structure of GFP in detail and then altered the sequence of the GFP chromophore and surrounding amino acids. By mutating them he was able to manipulate the excitation and emission wavelengths of the protein and so produce cyan, yellow, blue, and a veritable spectrum of fluorescent proteins. These then allowed biologists to track multiple proteins and dynamic processes simultaneously.

Nanoscopes

The light microscope is undoubtedly an excellent tool for inspecting the minute worlds that lie beyond our vision. Unfortunately, physics sets a limit on what can be resolved with visible light. Ernst Abbe described this 'diffraction limit' in 1873 when he showed that it is impossible to resolve two objects that are closer together than half the wavelength of the light being used to observe them. Since visible light has wavelengths ranging from 400 to 700 nanometres, this means that, practically speaking, light microscopy is limited to studying objects no smaller than 0.2 micrometres, about the size of a complex virus such as variola (which causes smallpox). Meanwhile, biological molecules, which are typically ten to a hundred times smaller than this, are beyond the reach of traditional light microscopy.

The most obvious way to get around this problem is to interrogate biomolecules using something with a much shorter wavelength. Electrons, with their wavelength in the order of 1 nanometre, work well, and form the basis of electron microscopy. This was the approach used by Cecil Hall, who, in 1956, utilized a transmission electron microscope to reveal long fibres of DNA carefully stretched out between polystyrene beads (Figure 26).

As amazing as it must have been to see single molecules of DNA for the first time, these images provide little useful information about DNA. After all they came three years after Franklin, Watson, Crick, Wilkins, and Gosling had published their crystallographic DNA structure. History clearly shows which study had the greater impact. Electron microscopy also comes with its own physical limitations. Biological molecules, consisting of fairly light atoms, are not very good subjects for transmission

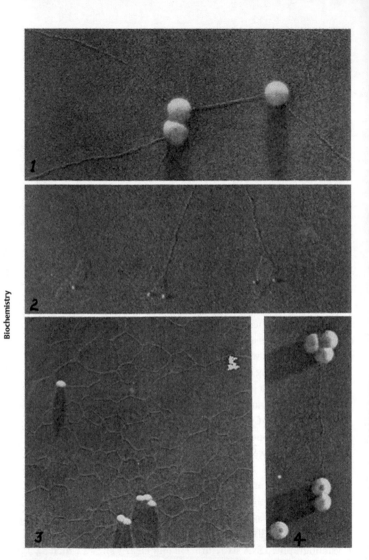

26. Cecil Hall's first electron micrography image of a single molecule of DNA stretched between two polystyrene beads.

electron microscopy. To resolve them at all they need to be coated with a heavy metal, such as gold or platinum. Samples need to be mounted on charged surfaces and dried—circumstances bearing little resemblance to a biologically relevant environment. Biophysicists therefore never really gave up on light microscopy, and instead tried to come up with ways to circumvent the diffraction limit. The particular efforts of Eric Betzig, Stefan Hell, and William Moerner were recognized when, in 2014, they were awarded the Nobel prize in Chemistry 'for the development of super-resolved fluorescence microscopy'.

These ultra high resolution microscopes, or nanoscopes as they are sometimes called, are developments of a very well utilized tool called fluorescence microscopy. For example, a scientist might be interested in a protein she suspects is involved in DNA repair. She labels the protein using a GFP tag that emits green light, and as a result the chromosomes in the nucleus of the cell light up green, supporting her hypothesis. However, because of Abbe's diffraction limit she'd never be able to get much more resolution than that.

That was until Betzig, Hell, and Moerner (amongst others, although just this trio were recognized for the Nobel prize) demonstrated that light microscopy could resolve single molecules under the right conditions. Betzig built on the work of Moerner, who had found a loophole in Abbe's limit. Moerner had shown that the diffraction limit did not preclude observation of single molecules, as long as they were sufficiently far apart. Two objects closer than 0.2 micrometres produce a single blur wider than 0.2 micrometres, whilst a single fluorescing molecule produced an image that blurred over just 0.2 micrometres. So, if you know you only have one molecule in your image then the chances are that molecule is situated at the middle of the blur. Then with a clever bit of image processing, using probability theory, you could reconstruct a sharp image. The trick is to work out how to get signals from just a few molecules at a time, so that their signals do not overlap into that wide blur. The solution came from more of Moerner's work on GFP. If GFP is excited with blue light at 488 nanometres it emits green light at 509 nanometres, but it does not do so indefinitely; the emission fades over time. Moerner found that the emission could be reactivated by illuminating the GFP with violet light at 405 nanometres.

Betzig combined this knowledge to build an apparatus that produced astonishing images of single molecules within cells. First a membrane protein located on lysosomes was labelled with a fluorescent protein. Next the cells were bathed with blue light, until all the protein stopped fluorescing. Then proteins were reactivated with very weak violet light. The key to the technique is that the reactivating beam is so weak only a tiny fraction of the proteins resume their fluorescence. This means that each emitting protein is probably more than 0.2 micrometres from its nearest emitting neighbour, and so they can each be individually resolved. By repeating the process multiple times, and superimposing the resulting images, it becomes possible to create super-resolution light 'nanoscope' images revealing structures and interactions well below Abbe's diffraction limit.

Since Betzig first demonstrated this technique in 2006 numerous other versions of super-resolution microscopy have been developed (including the methodology developed by Hell), but underlying them all is the need to keep the majority of molecules dark in any given frame of the image. This incredible resolution boost does, however, come with a major limitation. It is not possible to see these amazing images through a microscope—they are composites of many thousands of individual exposures. The image in Figure 27 took most of a day to construct.

Nevertheless, together these techniques are producing hitherto unimaginable images and insights into the interactions of biomolecules. They have allowed us to visualize individual pores on the surface of cells, single DNA polymerase enzymes working within the replisome of a living cell, and the arrangement of

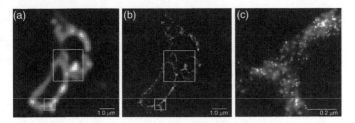

27. Image (a) was taken using conventional microscopy; image (b) is the same area, but now resolved using Betzig's nanoscope; and image (c) is a further expansion of the area within the marked square. (Note the scale bar of 0.2 micrometres, which indicates the size of the normal diffraction limit.)

proteins on the surface of nerve cells. But the most compelling revelations may come from work on one of the most fundamental processes of life: cell division.

Super high resolution insights into cell division

Cytokinesis, the process of cell division, is an astonishing mechanical feat, comparable to cutting an inflated balloon in half without popping it. The analogy is particularly apt, given that some bacteria maintain an osmotic pressure of over 20 atmospheres. Every time a cell divides, it has to find its midpoint, start building a dividing wall (referred to as a septum), and oversee scission of the membranes, all the while making sure each daughter cell receives an equal share of the cellular machinery and an exact copy of the genome. This is made all the more remarkable when you consider that some bacteria carry out this task every twenty minutes.

Eukaryotes and prokaryotes approach the cytokinesis challenge in different ways, so for the sake of brevity, and to illustrate the role that super-resolution microscopy has played in understanding the mechanisms, I will focus on one particular pivotal protein in prokaryotes.

A key player in bacterial cytokinesis was identified when a particular gene was knocked out. The resultant mutant bacteria continued to grow well past the point at which they would normally divide and instead formed long filamentous cells. The protein coded by that gene was given the name FtsZ (for **F**ilamenting **t**emperature-**s**ensitive mutant **Z**). The first hint at FtsZ's role came in the early 1990s from an electron microscopy study. Antibodies were raised to FtsZ and tagged with gold particles, which meant that the gold label and hence the protein could be spotted with an electron microscope. This clearly showed a band of FtsZ wrapping around the centre of a dividing cell. Later other studies, using fluorescence labels and optical microscopy on live cells, seemed to corroborate what had become known as the Z-ring, banded around the equator of a cell. These observations led to an assumption that a septum formed as the Z-ring anchored itself to the inside of the cell membrane and then contracted. This dragged the membrane in, pinching it towards the centre of the cell until it formed a complete wall between the nascent daughter cells.

The simplicity and elegance of the Z-ring acting like a belt tightening around the waist of the cell made it a rather compelling theory. And so it stood for some time, until holes were, quite literally, found in the Z-ring theory. As super-resolution imaging techniques developed and were applied to cytokinesis they consistently showed large gaps in Z-rings, meaning that FtsZ could not be forming a constricting band around the middle of cells.

About the same time as the constricting Z-ring hypothesis was unravelling, a study of single molecules of FtsZ showed up some unexpected behaviour. The protein was observed forming polymers which then appeared to move across the surface of lipid membranes. This was particularly surprising as there are no known motor proteins within bacteria, so it was difficult to imagine what was driving the movement of the FtsZ structures.

Further observations revealed that individual FtsZ molecules do not actually move at all. Instead the structure only appears to move due to a mechanism referred to as 'treadmilling'. This occurs when a filament grows by the addition of a protein at one end, while at the same time a protein drops off the apparently trailing end, so giving the impression of movement. More tracking of FtsZ, and proteins associated with it, revealed that the purpose of the complex was not to apply a constricting force on the cell and pinch off daughters. Instead it seems the fragmented Z-ring actually consists of FtsZ polymers treadmilling around the equator of the cell carrying a cargo of proteins that manufacture membrane while simultaneously building the septum wall that eventually separates the daughter cells.

Cytoskeletons and motor proteins

Cells are much more than simple bags of cytoplasm, organelles, and biomolecules inside a membrane. All these constituents are organized around a complex and dynamic three-dimensional structure of interlinking filaments known as the cytoskeleton. This provides mechanical support to the cell, but also acts as an intracellular highway travelled by motor proteins and their cargoes. The cytoskeletons of eukaryotes are made up of three types of filaments. Intermediate filaments are the least dynamic and are primarily for physical support; these are constructed from a variety of proteins depending on the filaments' function. The other two are micro-tubules (formed from polymerized tubulin sub-units) and actin micro-filaments.

Myosin motor proteins use the actin filaments to rearrange the positions of organelles within a cell (actin and myosin are also the main components of muscle tissue). Meanwhile, the micro-tubule network is used for transporting much smaller cargoes across long distances. Micro-tubules (and micro-filaments) also have a polarity, and the two motor proteins that traverse them hold to a one-way rule, only ever moving in one direction along the tubule.

Dyneins carry cargo from the periphery of the cell towards the nucleus at its centre, while kinesins perform the return journey.

The complexity of the cytoskeletal systems, their cargoes, and the roles they play in organizing cells and cell division has thrown up a host of questions, and answering them proves to be particularly fertile ground for single-molecule biophysicists.

One question that immediately springs to mind, given that kinesins and dyneins only move one way along micro-tubules, is how do they get back to their starting positions? There are three possible, obvious answers to this question: Once a protein reaches its destination, it could degrade, its parts being recycled back into the cell's metabolic systems. The second model proposes that once the motor reaches its destination it simply falls off the highway and finds its way back to where it is needed by passive diffusion. A third is that kinesins hitch rides with dyneins, and vice versa. Given that there are numerous variations of kinesin and dynein in a given cell and even more in the whole of the biological world, these models are not mutually exclusive, and indeed there is evidence for all three, but some of the most intriguing studies focus on the third possibility and an apparent tug-of-war between the motor proteins.

One particularly nice study reconstructed a kinesin and dynein transport system from yeast and followed the movement of the proteins using fluorescent tags and single-molecule microscopy. The scientists observed the kinesin and dynein motors move in opposite directions along the micro-tubules, exactly as expected. Then when a protein called Lis1 was added to the mix, it coupled the motor proteins together. This created a tug-of-war situation with the proteins battling against each other to reach opposite ends of the micro-tubules. The stalemate was resolved with the help of two other micro-tubule associating proteins, which appeared to give kinesin some extra 'grip'. This additional traction helped kinesin overcome dynein and win the tug-of-war. The rather elegant study suggests that the motor proteins may well be recycled by each other, with the direction of travel being mediated by other minor proteins.

Optical traps and tweezers

So far, during this chapter, I have concentrated on the two most productive techniques for watching single molecules at work within cells. I have deliberately overlooked other single-molecule techniques because they are normally deployed *in vitro*. However, one method, using what are known as optical tweezers, is providing a means for us to transition from *in vitro* to *in vivo* study.

Tractor beams and their apparent capacity to use electromagnetic radiation to move physical objects have been a mainstay of science fiction. However, it now turns out that the idea may no longer be confined to fiction. James Clerk Maxwell had presented a theoretical basis for light exerting physical pressure on an object as far back as 1873. However, with the forces involved being extremely weak, it took a further 130 years before Maxwell's predictions could be tested and verified. In the 1960s, laser technology emerged, which, at the time, was described as 'a solution looking for a problem'. It turned out to have manifold uses: everything from pointing at slides in a presentation, to scanning bar codes, and to manipulating single biological molecules. And in time it transpired that lasers, with their highly focused and intense beams, at last provided the means for pressure to be exerted by light, and to be used in ways that had, hitherto, only been available to James T. Kirk.

The pioneer in the field of optical manipulation of objects was Arthur Ashkin (yet another Nobel Laureate, this time in physics in 2018). In one of his first breakthroughs, he managed to levitate a particle by balancing the upward force of laser light against the downward pull of gravity. Later he showed that the combined

A motor molecule walks inside the light trap

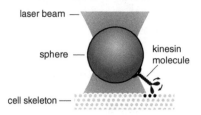
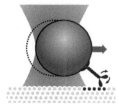

The kinesin molecule attaches to a small sphere held by the optical tweezers.

Kinesin marches away along the cell skeleton. It pulls the sphere, making it possible to measure the kinesin's stepwise motion.

28. Optical tweezers tracking molecular motors.

forces of a focused beam drew a particle to its focal point, with enough force to overcome Brownian motion. This effectively held an object at a point in space defined by the position of the laser beam, creating an optical trap. Consequently, a laser could be used as 'optical tweezers' (Figure 28) to 'pick up' and move particles by changing the focal point of the laser. A further development to the technique allowed the forces being exerted on the optical tweezers to be measured. This meant it became possible to quantify the tug of an individual molecular motor.

In 1993, optical tweezers were used to directly observe how a kinesin molecule moves. Karel Svoboda, Steven Block, and colleagues decorated silica beads with kinesin and then used their optical tweezers to deposit them onto immobilized micro-tubules. This allowed Svoboda et al. to manipulate the beads, and measure the force produced by kinesin. Each one exerted a force of just 5 piconewtons, and moved in discrete 8-nanometre steps along the micro-tubule track.

Over the intervening years since Svoboda et al.'s pioneering work, optical tweezers have been used to monitor a host of biochemical processes, including the movement of other molecular motors, proteins working their way along DNA, and forces involved in folding and unfolding proteins. More recently optical traps have been developed to the point where they can now be used to follow molecules moving in a similar way within cells.

The main challenge to *in vivo* optical trapping was overcoming the background noise and light scattering generated by the rest of the cell. But once this technical hurdle was cleared, the technique remains fundamentally the same.

It transpires that cellular cargoes frequently have both dynein and kinesin motors attached to them, which has implications for how cells run their transport networks. And *in vivo* optical trapping has shed light on the dynein/kinesin tug-of-war. One of the first studies of this type involved the absorption by cells of beads adorned with kinesin and dynein (via a natural process called endocytosis), which were then manipulated with optical tweezers. The study found that *in vivo* kinesin and dynein move with forces very much as they do *in vitro*. However, when the two motors pull on the same particles, the interactions become more complicated. When dynein is dominant it carries its cargo along the micro-tubule without the kinesin having any apparent effect. Presumably the kinesin acts as a passive partner, detaching from the micro-tubule and being carried along. However, when kinesin is the driver the dynein appears to present some resistance, as it moves backwards along the micro-tubule. The assumption is that this interplay may help to deliver cargoes to precise locations within cells.

I started this chapter by describing the benefits of single-molecule studies from a biochemist's perspective, which focuses on mapping the network of interactions between molecules. However, I have also made reference to biophysics, and one reason for doing so is that the techniques described in this chapter utilize so many physical principles. With that we also start to draw in the physicist's perspective of biomolecules, and physicists like to quantify interaction in terms of forces involved. So many single-molecule techniques don't just show us how the molecules interact but also the forces involved in their interactions. Through these we now know that a kinesin motor exerts 5 piconewtons; that a protein can be unfolded by pulling on its ends with about 12 piconewtons; and that DNA polymerase converts the hydrolysis of ATP into over 50 piconewtons of force. These insights are altering the way we think about microscopic biomaterials and making us recognize that force is a factor that shapes biological processes.

Chapter 8
Biotechnology and synthetic biology

Most of this book has dealt with 'natural' biochemistry. But as we know more about the biochemical world we start to consider how we might manipulate it for our own benefit. This has led to a field that is frequently called biotechnology and more recently synthetic biology. What this encompasses is still being defined, although the Royal Society of London makes a good stab at it, describing synthetic biology as 'design and construction of novel artificial biological pathways, organisms or devices or redesign of natural biological systems'.

The micro-scale of synthetic biology, the pathways, and the organisms included in this definition clearly draw on knowledge from biochemistry. But its philosophy is more closely aligned to the principles of engineering than to those of pure science. It is now common to see the engineering development cycle of 'design-build-test-improve' applied at the micro-scale along with a desire for modular and standardized outputs. We have already seen the start of these in the examples already discussed, such as green fluorescent protein, PCR, and DNA sequencing technologies. Since those techniques emerged in the 1980s great strides have been made, with the creation of synthetic genomes, recoding of genetic codes, nanomachinery engineered from proteins, and much more besides. There is no better gauge for how far things have come than the change in the cost of DNA sequencing. The Human Genome Project completed its goal of sequencing a human genome in 2003, at a cost of $2.7 billion. Fewer than twenty years later there are companies who will do the same work for less than $1,000 per genome. There is a compelling correlation between the falling cost in an ever-expanding computer processing capability, and the ubiquitous use of computers and decreasing costs in biotechnology. So in this final chapter we'll take a look at some more examples of synthetic biochemistry, the potential they hold, and briefly consider what this new field might herald.

Synthetic organisms and genomes

If there is a poster boy for synthetic biology then it is probably Craig Venter. Not content with leading a private team that sequenced a human genome (his own, as it turned out), three years ahead of when the public funded projects were expected to complete the task, he also set his sights on creating synthetic organisms.

Venter's team, working at the J. Craig Venter Institute, approached the challenge by starting with a very simple organism, the bacterium *Mycoplasma mycoides*. With its genome of about 1,000 genes coded by about one million base pairs (compare that to almost five million for *E. coli* and three billion for humans) *M. mycoides* seemed like a manageable starting point on which to base an engineered lifeform. But before they could embark on that venture, they first needed to know whether a completely artificially manufactured genome could even 'boot up' a cell. So they redesigned the genome containing all the genes in the original *M. mycoides*, plus some 'watermark' sequences, which were included, according to their paper in *Science*, in order to differentiate the synthetic from the natural genome. (And, probably just for the fun of it, one watermark contained an additional DNA sequence that, when translated into amino acid single-letter codes, spelled out 'CRAIGVENTER'.) As the project developed, more elaborately encoded watermarks were included. One was a tribute to the great Richard Feynman, quoting the last words left on his blackboard after he died, 'What I cannot build, I cannot understand'. The quote has become almost a mantra of synthetic biologists.

By 2010, the synthetic genome JCVI-syn1.0 (a software-like name, after the institute in which it was created) had been constructed and was ready for transplanting into a DNA-free 'shell' derived from another bacterium, *Mycoplasma capricolum*. The newly inserted genome immediately took charge of its host's cellular machinery and together they became a completely viable self-replicating bacterium. As remarkable as this is, it is debatable whether the new system actually constitutes a synthetic organism, since the genome was largely a copy of the natural one.

The next stage of the project went some way to removing this objection. The team set about systematically paring down the genome to determine the minimum number of genes required to run an organism. By 2016 the genome, upgraded to JCVI-syn3.0, had been trimmed by more than half and now contained just 473 genes. Yet still the host bacteria were viable. And since this genome's design constitutes a reorganization and artificial selection of its contents (even if no recoding of the actual genes took place) it is safer to say that *Mycoplasma laboratorium* (sometimes called Synthia), now running the new artificial 'software', can truly be considered a synthetic biological lifeform. Interestingly the team had also tried to totally reorganize the layout of Synthia's genes, but found that for most of the genome this wasn't possible. This failure tells us something important: that the position of genes relative to each other is a critical, if yet unexplained, factor. And one that needs deciphering if we are to determine the rules governing genome engineering and design.

It is easy to become distracted by the magnitude of these sorts of accomplishments and forget that the ultimate aim of Venter and other synthetic biology researchers is to create organisms engineered to solve real-world problems. Venter's approach is to cut back the complexity of life to its bare minimum and then use the result as a foundation on which to build organisms designed to produce biofuels, pharmaceuticals, and materials. Others have taken a less dramatic approach and sought to tweak the biochemical pathways in existing organisms.

An excellent example comes from a paper published whilst I was writing this chapter. It describes how the bacterium *E. coli* can be engineered to absorb carbon dioxide. Normally, *E. coli* is a heterotroph, meaning it needs to consume other sources of organic carbon to grow. Conversely, plants and some bacteria are autotrophs; they are capable of fixing carbon dioxide from the atmosphere and converting it to organic biomass. We saw how they achieve this using photosystems and the Calvin cycle earlier in the book.

E. coli's transformation from heterotroph to autotroph required three steps. First, *E. coli* has no enzymes capable of fixing carbon dioxide, so genes coding for the enzymes within the Calvin cycle, sourced from the autotrophic bacterium *Pseudomonas*, were inserted into *E. coli*. Next, three of the host's own genes central to carbon metabolism were disabled, leaving bacterial growth dependent on the non-native enzymes. At this point, there was a problem because the *Pseudomonas* enzymes, expressed by the new genes, are alien to *E. coli* and so failed to integrate with the host metabolism. This isn't surprising since it had evolved for heterotrophic growth: expecting it to immediately make use of the new energy source would be like putting diesel in a petrol engine and being surprised when it doesn't work.

But unlike mechanical engines organisms can adapt to their circumstances. In order to push the *E. coli* to switch to its new fuel source evolutionary pressure was exerted on the redesigned metabolism. The modified bacteria were kept in a constant near-starvation state, but in an enriched carbon dioxide environment. Under these conditions the only source of carbon for accumulating biomass was the carbon dioxide, and the only way to access it was via the non-native Calvin cycle enzymes. After 200 days in these conditions the bacteria had accumulated eleven mutations, which allowed the alien enzymes to integrate with the *E. coli* metabolic pathways. The result was a new variant of *E. coli*, tailored through a combination of engineering and directed evolution, that was completely reliant on carbon dioxide for its biomass.

Gene editing

Proteins are life's molecular machines, and so they are also the primary focus of biomolecular engineering. But manipulating them directly is tricky. Instead it is much simpler to edit and move the genes that code for proteins.

For many years the main way to move and manipulate genes was with a class of protein called restriction enzymes. These are derived from bacteria, where they play a role in defence against phage (viruses that attack bacteria) infection, by cutting up the DNA of invading organisms. To achieve this, whilst avoiding cutting their own DNA, the restriction enzymes generally recognize specific palindromic DNA sequences. For example, one of the most widely used restriction enzymes, known as HindIII (pronounced 'hin-dee-three'), recognizes the sequence AAGCTT and the complementary sequence of TTCGAA (which is the first sequence in reverse, hence it is palindromic). HindIII cuts between the As on both DNA strands, leaving overhangs of single-stranded DNA. Meanwhile another commonly used example, EcoRI (pronounced 'eco-ar-one'), cleaves after the G in the DNA sequence GAATTC.

This feature proves to be extremely useful for genetic manipulation, particularly of bacteria. Bacteria frequently contain small genetic structures that can replicate independently of the chromosomal DNA. These structures, known as plasmids, are useful vectors for carrying genetic material into a bacterium. And this is exactly what bacteria use them for. For example, bacteria share antibiotic resistance genes by transferring plasmids to one another. This propensity to transfer plasmids is utilized by scientists to manipulate bacteria.

These days, artificially constructed plasmids can be purchased with a region, known as a multiple cloning site (MCS), containing a series of restriction enzyme digestion sequences (Figure 29). This MCS is positioned just upstream of promotors. This means that when a gene is inserted into the MCS the transcription machinery kicks into action and the bacterium ends up producing the protein coded by the gene. It is via this process that bacteria have been persuaded to make a range of useful proteins such as insulin, rennet, and growth hormones.

To insert a gene into the plasmid the desirable restriction enzyme site is first identified. We now know of literally hundreds of

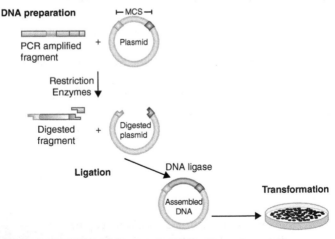

29. **Gene cloning strategy using PCR, plasmids, and restriction enzymes.**

restriction enzymes, most of which are commercially available. Next PCR is used to make copies of the gene of interest, and the restriction enzyme sites are included within the DNA primers. These then become incorporated into the product of the PCR reactions. The plasmids and the gene, now multiplied by PCR, are then cut with the restriction enzymes. The cut ends of all these DNA fragments will have overhangs, all of which contain complementary sequences, helping them to attach to one another. The final step, of annealing the pieces of DNA, is achieved by addition of DNA ligase. And then a bacteria is persuaded (in a process called transformation) to take up the plasmid containing the new gene.

This restriction enzyme method to manipulate DNA has been used for decades, but it is rather laborious and limited. Restriction enzymes are undoubtedly extremely useful, but it's a bit like having hundreds of 'cut and paste' options on your word processor, with each one being specific only for a particular word. A more versatile cut and paste tool would be significantly more valuable.

CRISPR

In recent years just such a powerful gene editing technique, known as CRISPR (**C**lustered, **R**egularly **I**nterspaced, **S**hort **P**alindromic **R**epeats), has emerged. Unlike the restriction enzymes, it allows for 'cut and paste' and 'find and replace' functions at more or less any portion of a DNA sequence within a cell's genome. The implications are huge. It has already been used to remove HIV genes from infected human cells, delete mutated genes that cause disease, turn on the gene for the foetal version of haemoglobin, treat sufferers of sickle cell disorder, and modify animal organs with the aim of making them viable for transplantation into humans.

CRISPR's story starts in 1987 when Yoshizumi Ishino and colleagues noticed repeating palindromic DNA sequences, of

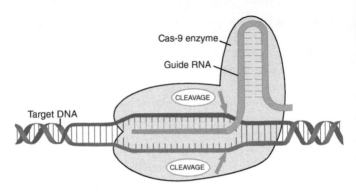

Cas-9 enzyme

Guide RNA

CLEAVAGE

Target DNA

CLEAVAGE

30. The CRISPR-CAS-9 system.

about thirty bases, in *E. coli* bacteria. Next to the palindromes there were highly variable sequences with little discernible pattern. The team had no idea what the strange feature in the genome was for, but they published the finding and then didn't think much of it. Over the next decade or so similar patterns cropped up in other prokaryotic systems and so interest in them grew. The key to elucidating their function came when the variable sequence regions were compared to other known DNA sequences and found to be homologous to bacteriophage genes. And so the theory began to emerge that these CRISPR sequences formed part of some sort of adaptive immune system. This was tested when a particular phage DNA was inserted into the variable region, and sure enough the bacteria containing this newly identified CRISPR region became resistant to that phage.

Next scientists spotted genes coding for helicases and nucleases (proteins that cut nucleic acids) appearing near the CRISPR regions (and these became known as CAS-CRISPR **As**sociated-genes; Figure 30). It transpires that the CRISPR palindromic sequences, when transcribed into RNA, form hairpin secondary structures. One of the CAS proteins (CAS-9) recognizes and binds to the shaft of the hairpin and presents the variable sequence in a way that leaves it free to bind to complementary viral DNA or RNA that the phage had injected into the cell. Hence the RNA sequence guides the CAS protein to the phage genes, and so it became known as guide RNA (gRNA). When the gRNA binds to the target viral genes, it puts the CAS's nuclease activate site in proximity to the invading viral genetic material, allowing the protein to cut clean through both strands of the viral DNA, so stopping the infection in its tracks. This, of course, only works if the bacterium has a record of the viral genetic material in one of the spacer regions. To make that record another class of CAS protein recognizes incoming DNA, cuts out a chunk and inserts it between palindromic sequences. In so doing it retains a 'memory' of the infection, allowing the bacteria to respond more quickly in the future.

Once the CRISPR and CAS mechanisms became apparent researchers, and most notably Jennifer Doudna, Emmanuelle Charpentier (who together received the 2020 Nobel prize for Chemistry), and Virginijus Šikšnys, realized that the system could be programmed and engineered to cut any DNA sequence, merely by replacing the natural spacer sequence with one that was complementary to a given gene. This allowing them to precisely target and disable any gene. The technique is therefore perfect for inactivating a faulty gene. But what if you want to replace that gene with a functioning sequence? To achieve this, scientists adapted another naturally occurring biochemical process called homology directed repair.

A break in a single strand of DNA is easily fixed: the complementary strand acts as a splint holding everything together whilst a ligase enzyme stitches up the break. Double stranded breaks, frequently caused by environmental factors such as ultraviolet light, chemical agents, or CRISPR, are much more troublesome. The ends could simply be reannealed, but what if there has been more than one break and a section of DNA is missing? Then the break could cause degradation of a chromosome, genomic rearrangements, and inactive or unregulated genes leading to cell death, or, worse, cancerous cells. What the cell needs is a spare copy of the missing sequence on which it can base the repair, and that can usually be found on another chromosome. The cellular machinery recognizes homologous regions and then replicates any missing sections based on the intact chromosome, before slotting them into the break.

This means that if we want to insert a gene into a genome, we simply adapt the CRISPR system to cut out a section of DNA, and then swamp the cell with copies of the gene we want to insert, sandwiched between regions homologous to the sequences on either side of the break. The homology directed repair system will then most probably use the artificially injected DNA to fix the CRISPR induced break.

Designer babies

For all the promising applications that synthetic biology affords, from synthetic organisms that might create sustainable fuels to editing out faulty genes, there are serious ethical issues we need to grapple with and darker applications of these technologies.

In November 2018, He Jiankui of the Southern University of Science and Technology in Shenzhen, China, shocked the world by announcing the birth of gene edited twin girls. His aim was to make them immune to HIV infection. And to achieve this he fertilized eggs from a consenting donor, using her partner's sperm.

Then He used CRISPR to disrupt a gene called CCR5 in the embryos. Normally this gene expresses a receptor protein used by HIV to latch on to and infect cells. The donor couple agreed to have the embryos implanted and both embryos developed normally, resulting in live births.

He's experiments—and they were very much experiments—were highly controversial, not least because there was no medical need for the procedure. The father was HIV positive, but tried and tested techniques are already available to remove the HIV virus from sperm samples before *in vitro* fertilization (IVF) treatment, so eliminating the risk of HIV infecting mothers and their children. Furthermore, the CCR5 gene plays a positive role in the immune system, and by removing it He's actions may have a detrimental effect on the girls' health. But even worse than this, by meddling with the embryos, He had altered the germ line, and should the girls have families, their children will also carry the edited genes, and their unknown consequences. He was subsequently convicted of 'illegal medical practices' and sentenced to three years in prison.

Designer bugs

In 2002, whilst work on JCVI-syn1.0 was just beginning, Dr Eckard Wimmer published a groundbreaking paper describing how his team had created a functional replica of a pathogen virus. Wimmer started by downloading the publicly available genome sequence of poliovirus. Purchasing short stretches of DNA from one of a multitude of companies that will provide sequences to order, Wimmer and his team then pieced the short sequences together until they had recreated the whole 7,741-base genome. Polio has an RNA genome, so the DNA had to be transcribed back into RNA. When the synthetic genome was inserted into a cell line the cells produced viral particles which caused polio in mice.

As Wimmer himself put it, 'You no longer need the real thing in order to make the virus and propagate it.' All you need is the genome sequence, access to DNA sequences that you can order online to receive in the post, and a couple of skilled technicians.

Consider, then, that since 2002 biotechnological capabilities have expanded considerably. In 2005 scientists at the Center for

Disease Control, in the USA, reconstructed the 1918 Spanish Flu virus and in 2018 a Canadian team created horsepox, a relative of smallpox, from mail-order DNA. Horsepox poses no risk to human health, but by demonstrating how to create a complex virus (its genome is some twenty times bigger than polio), their developments take us further towards creating synthetic smallpox.

Another worrying development is gain-of-function research, whereby organisms are given additional 'abilities'. This was illustrated by the work of Ron Fouchier, who introduced just five mutations into the bird flu H5N1 virus, which converted it into an airborne virus capable of infecting mammals.

These developments make the accidental or nefarious release of dangerous pathogens a very real prospect. And as such there is an urgent need for regulatory frameworks to govern these emerging technologies. Some self-governance is already in place. Biotechnology firms now screen orders for DNA against known dangerous sequences. This was in response to an episode in 2006 when journalists at the *Guardian* newspaper looked up the sequence of smallpox and ordered a stretch of its DNA. But is this self-regulation enough? Dangerous pathogens are kept in tightly regulated, high security facilities. However, it took me about five minutes to find the complete sequence of the smallpox virus genome on a publicly accessible database. Previous ages have faced perils from the technology that they wrought, and we have had to grapple with the consequences. As we plunge into the biological age we now need to consider the implications of our knowledge and ask how we might control access to biological information. Maybe the blueprints that can be used to build dangerous organisms should be just as restricted as the organisms themselves?

In 2012, in her capacity as chief scientific advisor to the president of the European Union, Professor Dame Anne Glover addressed the Organisation for Economic Co-operation and Development

Global Forum on Biotechnology. She described how centuries are defined by scientific advances: the 19th century was the age of engineering, seeing development of society-altering technologies such as Stephenson's steam locomotive, Benz's motor car, and Marconi's radio. In the 20th century, Haber and Bosch triggered a population explosion through the development of a means to make ammonia and hence agricultural fertilizer on an industrial scale; Florey, Chain, and Fleming created a medicine that changed the way we deal with infections; Marie Curie developed the theory of radioactivity; and Tim Berners-Lee gave us the world wide web—as such the 20th century was the age of chemistry and physics. And, at the dawn of the 21st century, the human genome was sequenced, heralding an 'age of biology', with undoubtedly ever-expanding biochemistry-based technologies aimed at remodelling society (for good or ill), just as motor vehicles, antibiotics, and the internet have done in previous ages.

References

Preface

Schrödinger, E., 1967. *What Is Life?* Cambridge: Cambridge University Press.

Chapter 1: The roots of biochemistry

Hein, G., 1961. The Liebig-Pasteur controversy: Vitality without vitalism. *Journal of Chemical Education*, 38(12), p.614.

Kohler, R., 1971. The background to Eduard Buchner's discovery of cell-free fermentation. *Journal of the History of Biology*, 4(1), pp.35–61.

Hofmeister, F., 1902. Über Bau und Gruppierung der Eiweisskörper. *Ergebnisse der Physiologie*, 1(1), pp.759–802.

Fischer, E., 1906. Untersuchungen über Aminosäuren, Polypeptide und Proteïne. *Berichte der deutschen chemischen Gesellschaft*, 39(1), pp.530–610.

Sanger, F. and Tuppy, H., 1951. The amino-acid sequence in the phenylalanyl chain of insulin. 1: The identification of lower peptides from partial hydrolysates. *Biochemical Journal*, 49(4), pp.463–81.

Sanger, F. and Thompson, E., 1953. The amino-acid sequence in the glycyl chain of insulin. 1: The identification of lower peptides from partial hydrolysates. *Biochemical Journal*, 53(3), pp.353–66.

Pauling, L., 1951. *The Nature of the Chemical Bond and the Structure of Molecules and Crystals.* Ithaca, NY: Cornell University Press.

Pauling, L., Corey, R., and Branson, H., 1951. The structure of proteins: Two hydrogen-bonded helical configurations of the polypeptide chain. *Proceedings of the National Academy of Sciences*, 37(4), pp.205–11.

Pauling, L. and Corey, R., 1951. Configurations of polypeptide chains with favored orientations around single bonds: Two new pleated sheets. *Proceedings of the National Academy of Sciences*, 37(11), pp.729–40.

Soyfer, V., 2001. The consequences of political dictatorship for Russian science. *Nature Reviews Genetics*, 2(9), pp.723–9.

Avery, O., MacLeod, C., and McCarty, M., 1944. Studies on the chemical nature of the substance inducing transformation of pneumococcal types. *The Journal of Experimental Medicine*, 79(2), pp.137–58.

Hershey, A. and Chase, M., 1952. Independent functions of viral protein and nucleic acid in growth of bacteriophage. *The Journal of General Physiology*, 36(1), pp.39–56.

Pauling, L. and Corey, R., 1953. A proposed structure for the nucleic acids. *Proceedings of the National Academy of Sciences*, 39(2), pp.84–97.

Watson, J. and Crick, F., 1953. Molecular structure of nucleic acids: A structure for deoxyribose nucleic acid. *Nature*, 171(4356), pp.737–8.

Wilkins, M., Stokes, A., and Wilson, H., 1953. Molecular structure of nucleic acids: Molecular structure of deoxypentose nucleic acids. *Nature*, 171(4356), pp.738–40.

Franklin, R. and Gosling, R., 1953. Molecular configuration in sodium thymonucleate. *Nature*, 171(4356), pp.740–1.

Chapter 3: Proteins: nature's nano-machines

Berman, H., Battistuz, T., Bhat, T., Bluhm, W., Bourne, P., Burkhardt, K., Feng, Z., Gilliland, G., Iype, L., Jain, S., Fagan, P., Marvin, J., Padilla, D., Ravichandran, V., Schneider, B., Thanki, N., Weissig, H., Westbrook, J., and Zardecki, C., 2002. The Protein Data Bank. *Acta Crystallographica Section D Biological Crystallography*, 58(6), pp.899–907.

Koshland, D., 1995. The key–lock theory and the induced fit theory. *Angewandte Chemie International Edition in English*, 33(2324), pp.2375–8.

Michaelis, L. and Menten, M.L. 1913. Kinetik der Invertinwirkung. *Biochem. Z.* 49, pp.333–69.

Kendrew, J., Bodo, G., Dintzis, H., Parrish, R., Wyckoff, H., and Phillips, D., 1958. A three-dimensional model of the myoglobin molecule obtained by X-ray analysis. *Nature*, 181(4610), pp.662–6.

Anfinsen, C., Haber, E., Sela, M., and White, F., 1961. The kinetics of formation of native ribonuclease during oxidation of the reduced polypeptide chain. *Proceedings of the National Academy of Sciences*, 47(9), pp.1309–14.

Levinthal, C., 1969. How to fold graciously. *Mossbauer Spectroscopy in Biological Systems: Proceedings of a Meeting held at Allerton House, Monticello, Illinois*, 67(41), pp.22–4.

Foldingathome.org. 2020. *Folding@Home—Fighting Disease with a World Wide Distributed Super Computer*. Available at: https://foldingathome.org/ (accessed 10 March 2020).

Cameo3d.org. 2020. *CAMEO—Continuous Automated Model Evaluation—Welcome*. Available at: http://cameo3d.org/ (accessed 10 March 2020).

Bragg, W., 1913. The diffraction of short electromagnetic waves by a crystal. *Proceedings of the Cambridge Philosophical Society*, 17, pp.43–57.

Chapter 4: Nucleic acids: life's blueprints

Crick, F.H., 1958. On protein synthesis. In: F.K. Sanders, ed., *Symposia of the Society for Experimental Biology, Number XII: The Biological Replication of Macromolecules*. Cambridge: Cambridge University Press, pp.138–63.

Zaug, A., Been, M., and Cech, T., 1986. The Tetrahymena ribozyme acts like an RNA restriction endonuclease. *Nature*, 324(6096), pp.429–33.

Crick, F.H.C., Griffith, J.S., and Orgel, L.E., 1957. Code without commas. *Proceedings of the National Academy of Sciences*, 43(5), pp.416–21.

Nirenberg, M. and Matthaei, J., 1961. The dependence of cell-free protein synthesis in *E. coli* upon naturally occurring or synthetic polyribonucleotides. *Proceedings of the National Academy of Sciences*, 47(10), pp.1588–602.

Chapter 5: Powering a cell: bioenergetics

Krebs, H. and Johnson, W., 1937. Metabolism of ketonic acids in animal tissues. *Biochemical Journal*, 31(4), pp.645–60.

Mitchell, P., 1961. Coupling of phosphorylation to electron and hydrogen transfer by a chemi-osmotic type of mechanism. *Nature*, 191(4784), pp.144–8.

Chapter 6: Manufacturing and maintaining DNA

Meselson, M. and Stahl, F., 1958. The replication of DNA in Escherichia coli. *Proceedings of the National Academy of Sciences*, 44(7), pp.671–82.

Sanger, F., Nicklen, S., and Coulson, A., 1977. DNA sequencing with chain-terminating inhibitors. *Proceedings of the National Academy of Sciences*, 74(12), pp.5463–7.

Wade, N., 1998. *Scientist at Work: Kary Mullis; after the 'Eureka,' a Nobelist Drops Out*. Nytimes.com. Available at: https://www.nytimes.com/1998/09/15/science/scientist-at-work-kary-mullis-after-the-eureka-a-nobelist-drops-out.html (accessed 10 March 2020).

NobelPrize.org. 2020. *The Nobel Prize in Chemistry 1993*. Available at: https://www.nobelprize.org/prizes/chemistry/1993/mullis/lecture (accessed 10 March 2020).

Kleppe, K., Ohtsuka, E., Kleppe, R., Molineux, I., and Khorana, H., 1971. Studies on polynucleotides. *Journal of Molecular Biology*, 56(2), pp.341–61.

Barinaga, M., 1991. Biotech nightmare: Does Cetus own PCR? *Science*, 251(4995), pp.739–40.

Chapter 7: Following biochemistry within the cell

Neher, E. and Sakmann, B., 1976. Single-channel currents recorded from membrane of denervated frog muscle fibres. *Nature*, 260(5554), pp.799–802.

Chalfie, M., Tu, Y., Euskirchen, G., Ward, W.W., and Prasher, D.C., 1994. Green fluorescent protein as a marker for gene expression. *Science*, 263(5148), pp.802–5.

Abbe, E., 1873. Beiträge zur Theorie des Mikroskops und der mikroskopischen Wahrnehmung. *Archiv für Mikroskopische Anatomie*, 9(1), pp.413–68.

Hall, C., 1956. Method for the observation of macromolecules with the electron microscope illustrated with micrographs of DNA. *The Journal of Biophysical and Biochemical Cytology*, 2(5), pp.625–8.

Betzig, E., Patterson, G., Sougrat, R., Lindwasser, O., Olenych, S., Bonifacino, J., Davidson, M., Lippincott-Schwartz, J., and Hess, H., 2006. Imaging intracellular fluorescent proteins at nanometer resolution. *Science*, 313(5793), pp.1642–5.

Loose, M. and Mitchison, T., 2013. The bacterial cell division proteins FtsA and FtsZ self-organize into dynamic cytoskeletal patterns. *Nature Cell Biology*, 16(1), pp.38–46.

Belyy, V., Schlager, M., Foster, H., Reimer, A., Carter, A., and Yildiz, A., 2016. The mammalian dynein–dynactin complex is a strong opponent to kinesin in a tug-of-war competition. *Nature Cell Biology*, 18(9), pp.1018–24.

Roberts, A., Goodman, B., and Reck-Peterson, S., 2014. Reconstitution of dynein transport to the microtubule plus end by kinesin. *eLife*, 3.

Svoboda, K., Schmidt, C., Schnapp, B., and Block, S., 1993. Direct observation of kinesin stepping by optical trapping interferometry. *Nature*, 365(6448), pp.721–7.

Blehm, B., Schroer, T., Trybus, K., Chemla, Y. and Selvin, P., 2013. *In Vivo Optical Trapping Indicates Kinesin's Stall Force Is Reduced by Dynein during Intracellular Transport*. *Proceedings of the National Academy of Science*, 110(9), pp.3381–6.

Chapter 8: Biotechnology and synthetic biology

Royalsociety.org. n.d. *Synthetic Biology / Royal Society*. Available at: https://royalsociety.org/topics-policy/projects/synthetic-biology/ (accessed 10 March 2020).

Gibson, D., Glass, J., Lartigue, C., Noskov, V., Chuang, R., Algire, M., Benders, G., Montague, M., Ma, L., Moodie, M., Merryman, C., Vashee, S., Krishnakumar, R., Assad-Garcia, N., Andrews-Pfannkoch, C., Denisova, E., Young, L., Qi, Z., Segall-Shapiro, T., Calvey, C., Parmar, P., Hutchison, C., Smith, H., and Venter, J., 2010. Creation of a bacterial cell controlled by a chemically synthesized genome. *Science*, 329(5987), pp.52–6.

Hutchison, C., Chuang, R., Noskov, V., Assad-Garcia, N., Deerinck, T., Ellisman, M., Gill, J., Kannan, K., Karas, B., Ma, L., Pelletier, J., Qi, Z., Richter, R., Strychalski, E., Sun, L., Suzuki, Y., Tsvetanova, B., Wise, K., Smith, H., Glass, J., Merryman, C., Gibson, D., and

Venter, J., 2016. Design and synthesis of a minimal bacterial genome. *Science*, 351(6280), pp.aad6253.

Gleizer, S., Ben-Nissan, R., Bar-On, Y., Antonovsky, N., Noor, E., Zohar, Y., Jona, G., Krieger, E., Shamshoum, M., Bar-Even, A., and Milo, R., 2019. Conversion of *Escherichia coli* to generate all biomass carbon from CO_2. *Cell*, 179(6), pp.1255–63.e12.

Ishino, Y., Shinagawa, H., Makino, K., Amemura, M., and Nakata, A., 1987. Nucleotide sequence of the iap gene, responsible for alkaline phosphatase isozyme conversion in *Escherichia coli*, and identification of the gene product. *Journal of Bacteriology*, 169(12), pp.5429–33.

Jinek, M., Chylinski, K., Fonfara, I., Hauer, M., Doudna, J.A., and Charpentier, E. 2012. A programmable dual-RNA-guided DNA endonuclease in adaptive bacterial immunity. *Science* 337(6096), pp.816–21.

Cyranoski, D., 2019. The CRISPR-baby scandal: What's next for human gene-editing. *Nature*, 566(7745), pp.440–2.

Cyranoski, D., 2020. What CRISPR-baby prison sentences mean for research. *Nature*, 577(7789), pp.154–5.

Cello, J., 2002. Chemical synthesis of poliovirus cDNA: Generation of infectious virus in the absence of natural template. *Science*, 297(5583), pp.1016–18.

Pollack, A., 2002. *Traces of Terror: The Science. Scientists Create a Live Polio Virus*. Nytimes.com. Available at: https://www.nytimes.com/2002/07/12/us/traces-of-terror-the-science-scientists-create-a-live-polio-virus.html (accessed 10 March 2020).

Herfst, S., Schrauwen, E., Linster, M., Chutinimitkul, S., de Wit, E., Munster, V., Sorrell, E., Bestebroer, T., Burke, D., Smith, D., Rimmelzwaan, G., Osterhaus, A., and Fouchier, R., 2012. Airborne transmission of influenza A/H5N1 virus between ferrets. *Science*, 336(6088), pp.1534–41.

Randerson, J., 2006. Did anyone order smallpox? *The Guardian*. Available at: https://www.theguardian.com/science/2006/jun/23/weaponstechnology.guardianweekly (accessed 10 March 2020).

Further reading

For a broader history of protein science, see Tanford, C. and Reynolds, J., 2004. *Nature's Robots*. Oxford: Oxford University Press.

Watson, J., 1981. *The Double Helix*. London: Weidenfeld. James Watson's personal account of the discovery of the structure of DNA gives some fascinating insights into the science of the time and his appalling attitude to Rosalind Franklin.

There have been Nobel prizes aplenty for work described in this book, and many have excellent summaries on the Nobel prize website:

NobelPrize.org. 2020. *The Nobel Prize in Chemistry 1978*. Available at: https://www.nobelprize.org/prizes/chemistry/1978/summary/ (accessed 10 March 2020).

NobelPrize.org. 2020. *The Nobel Prize in Chemistry 1989*. Available at: https://www.nobelprize.org/prizes/chemistry/1989/cech/article/ (accessed 10 March 2020).

NobelPrize.org. 2020. *The Nobel Prize in Chemistry 1993*. Available at: https://www.nobelprize.org/prizes/chemistry/1993/press-release/ (accessed 10 March 2020).

NobelPrize.org. 2020. *The Nobel Prize in Chemistry 2008*. Available at: https://www.nobelprize.org/prizes/chemistry/2008/advanced-information/ (accessed 10 March 2020).

NobelPrize.org. 2020. *The Nobel Prize in Chemistry 2014*. Available at: https://www.nobelprize.org/prizes/chemistry/2014/popular-information/ (accessed 10 March 2020).

NobelPrize.org 2020. *The Nobel Prize in Chemistry 2020*. Available at: https://www.nobelprize.org/uploads/2020/10/advanced-chemistryprize2020.pdf (accessed 7 October 2020).

NobelPrize.org. 2020. *The Nobel Prize in Physics 2018*. Available at:
https://www.nobelprize.org/prizes/physics/2018/popular-
information/ (accessed 10 March 2020).

NobelPrize.org. 2020. *The Nobel Prize in Physiology or Medicine 1991*.
Available at: https://www.nobelprize.org/prizes/medicine/1991/
(accessed 10 March 2020).

The Biochemist ran an excellent special edition on 'Synthetic Biology',
2019, 41(3), pp.3–48.

Very Short Introductions are available on several topics mentioned in
this book including proteins, enzymes, synthetic biology, molecular
biology, and genes.

Index

For the benefit of digital users, indexed terms that span two pages (e.g., 52–53) may, on occasion, appear on only one of those pages.

Index

Biochemistry